I0505186

Purpose of this book is different from answering fascinating questions, which were covered in Time Matters, my previous book:

- **What is Time?** – Time is Quantum fluctuations, like Temperature is Brownian motion.

- **What is Gravity?** – Gravity is Time pressure: pressure from faster Quantum fluctuations in the direction of slower time.

- **How Matter and Galaxies appeared?** – From Time burning: there was no matter yet, but blobs of slow time collided at speeds that sometimes exceeded local speed limit.

- **Does the physics of Bob Lazar's story check out?** – It does!

- **What are UFOs? How can UFOs survive 1000×g acceleration? …**

The current book provides a clear foundation for physicists to discard Dark Science, which was cultivated for a century. Physics in this book is derived from classical mechanics in simple ways that even high school students can understand. The most advanced part used in this book is Newton–Laplace formula for a sound wave in any medium. High school students can safely take this amazingly concise formula on faith, and college students can learn how it is derived.

TABLE OF CONTENTS

★ – extended section, ★★ – new section, compared to Time Matters coverage.

ISBN: 9798329049114

1. METERING TIME

- **SECOND** is the time interval during which a light beam travels **299,792,458** meters
- Or, to make time unit definition more local: **NANOSECOND** (10^{-9} sec, a billionth of a **SECOND**) is the time interval during which a light beam travels about **30 cm** (about a **foot**).

$c = 299,792,458$ m/sec is the ==local speed of light== constant, by definition of **SECOND**.

Since Einstein, we know that the ==speed of time== varies from place to place and from time to time; it is not "absolute" as Newton claimed, and as our intuition suggests:

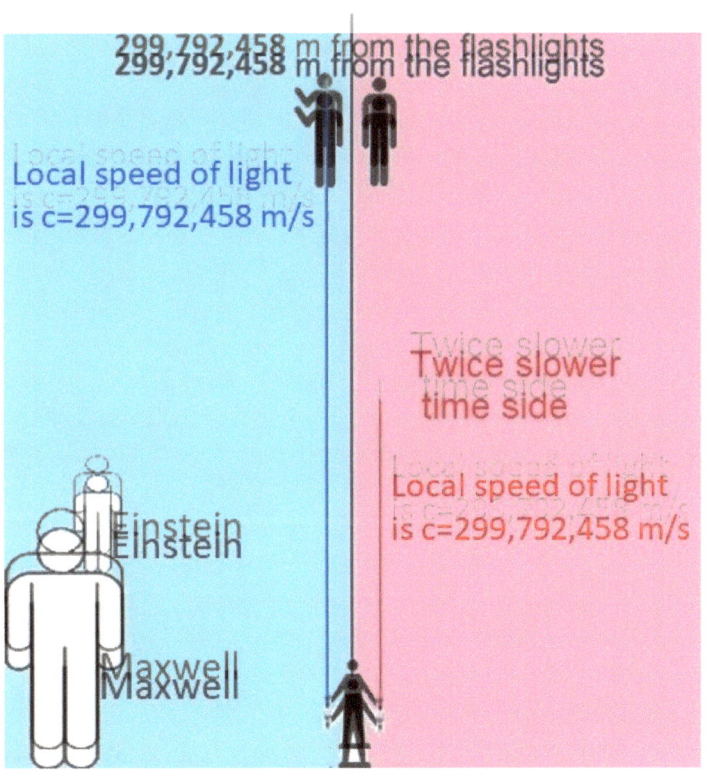

c is just a coefficient of proportionality (a constant ratio) between local speed of light and local speed of time. But even **NANOSECOND = 10^{-9} seconds** precision is not enough to detect time variability on the Earth; as illustrated on the front book cover, we need to work at 10^{-16} sec timescale. Even by the end of 20th century such precision was unachievable: 1999 article title "The Five Femtosecond Time Step Barrier" identifies 5×10^{-15} sec as "time step barrier". These numbers demonstrate why we don't have intuition of time flow variability. But nowadays physicists have tools and do experiments at 10^{-18} sec (which is called attosecond) timescale. 2023 Nobel Prize in Physics was awarded for a laser producing attosecond–apart pulses, and for its applications. In 10^{-18} sec, light (or photon) travels 3×10^{-10} meters. 10^{-10} m is a micro–unit called angstrom Å: it is the size of Hydrogen atom. 3×10^{-10} m is the size of larger elements, from iron to uranium atoms. The relation between the timescale at which physicists work nowadays and size of the atoms (which is the size of the electronic shell surrounding the atomic nucleus) is not coincidental: atomic clocks generate ticks by effects of electrons changing their "position/properties" in the atom. Only recently, physicists and engineers started working on utilizing atomic nucleus effects for time tracking, and that will give physicists thousand-times-smaller-than-attosecond time units to work with.

Now back to our basic drawing:

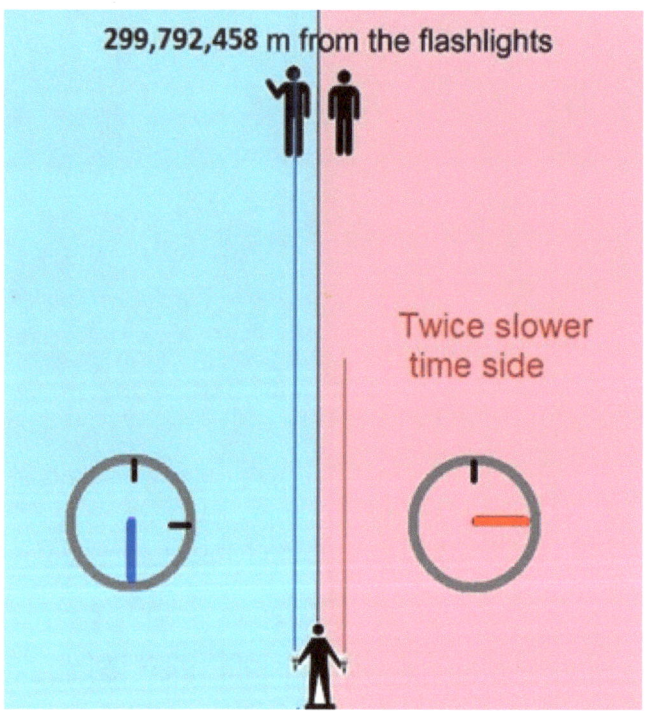

What is there besides the local speed of time and light? When the blue beam covered 299,792,458 m distance, the clock on the left/blue side showed 1 second passed, but the clock on the right/pink/twice-slower-time side showed only 1/2 seconds passed. Thus, from the right-side-time perspective, "apparent" speed of light on the left side is 299,792,458 m / (1/2 sec) = **2c**. And when the red beam traveled (299,792,458 / **2**)

meters on the right side, only 1/2 seconds passed on the right side, but a whole second passed on the left side, where clocks tick twice faster. Thus, from the left-side-time perspective, "apparent" speed of light on the right side is **c/2**. "Apparent" speed means speed measured in the time unit of an observer. Let say, an object moved distance **v** when on the left side 1 second passed, but on the right side only half a second passed. Then the "apparent" speed of this object is **v** from left-side-time perspective, and it is **v / (1/2) = 2v** from right-side-time perspective.

Einstein coined the "time dilation" term, and in this book, we denote time dilation factor as **D**, which means that local-to-the-object time is **D** times slower than time of the observer. For an observer in the blue area, **D = 2** in the pink area; and vice versa, for an observer in the pink area, **D = 1/2** in the blue area. Now we see that for an object moving at local velocity **v**, its apparent speed is **v/D** (because the observer's clock ticks **D** times faster than clocks in the area where the object is). The same applies to light: its apparent speed is **c/D**. That is true for both **D > 1** and **D < 1**. An apparent speed **v/D** is the same as local speed **v** when **D = 1** (when time speed is the same for an observer and for the observed object).

- **Velocity is a change in location per second**. For the observer, a second is **D**-times-faster/shorter than a local-to-the-object second. Thus, change during **D**-times shorter second apparently decreases by factor **D** (in comparison to the change per local-to-the-object second).
- Taking it further: **acceleration is change in velocity per second**. Observer's second is **D**-times-faster/shorter than local-to-the-object second. Thus, change in velocity per observer's second decreases again by factor **D**. Therefore, "apparent / observed" acceleration is **D²** smaller than local-to-the-object acceleration.

Let's remember this for the rest of the book as 3 simple formulas:

- Apparent velocity **v** = Local velocity **v / D**   (1.1)
- Apparent acceleration **a** = Local acceleration **a / D²**   (1.2)
- Apparent speed of light is **c/D**   (1.3)

2. OPTICAL EFFECTS OF TIME VARIABILITY

2.1. Color Change (Redshift, Blueshift)

Since Maxwell we know that light is an electromagnetic wave. Waves have frequencies = number of waves per second, which are expressed in the Hertz unit, which is 1 wave per second. Our naked eye can see light at about 400–800 terahertz range (terahertz is a trillion waves per second). Lower end of that frequency range is perceived by our brain as red light, and the higher end is viewed as blue light. Time dilation has an interesting effect on light frequency.

- Light received in a faster time area has B-times lower frequency than at its source, where time is slower by time dilation factor B. Astronomers say in such cases that received light was redshifted by factor $1+Z$, which is the same as B, because $B \equiv 1+Z$. The value Z is called "redshift" (which is the same as $B-1$, with $Z \equiv 0$ for "no redshift", which corresponds to $B \equiv 1$ when "no time dilation"). Redshift is just the effect of a different number of seconds passed at the source from the number of seconds passed for the receiver (number of waves does not change between the source and the receiver: waves do not appear or disappear in between).

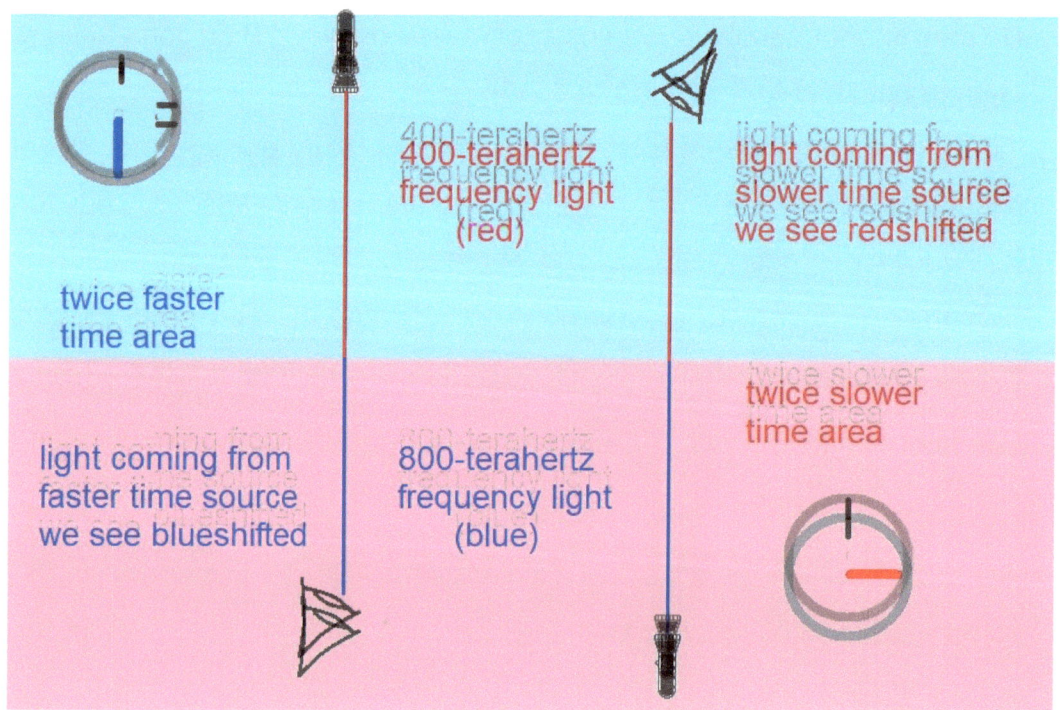

400-terahertz frequency light (red)

light coming from slower time source we see redshifted

twice faster time area

light coming from faster time source we see blueshifted

800-terahertz frequency light (blue)

twice slower time area

- When light is received in a slower time than the time at the light source, then $B<1$ and light is received at a frequency higher than it was at the source (as the drawing above explains). Astronomers say that such light is blueshifted.

Formula relating frequency of the light received f_R to its frequency at its source f_S is:

- $f_R = f_S / B;$ (2.1.1)

where B is time dilation at the source from the receiver perspective. A similar formula in terms of wavelength λ, which is inversely related to frequency f (by formula $f \times \lambda = c$), is:

- $\lambda_R = \lambda_S \times B.$ (2.1.2)

2.2. Refraction

When a light beam crosses borderline between areas with different time speeds, it is refracted by Snell's law with time dilation B as refractive index:

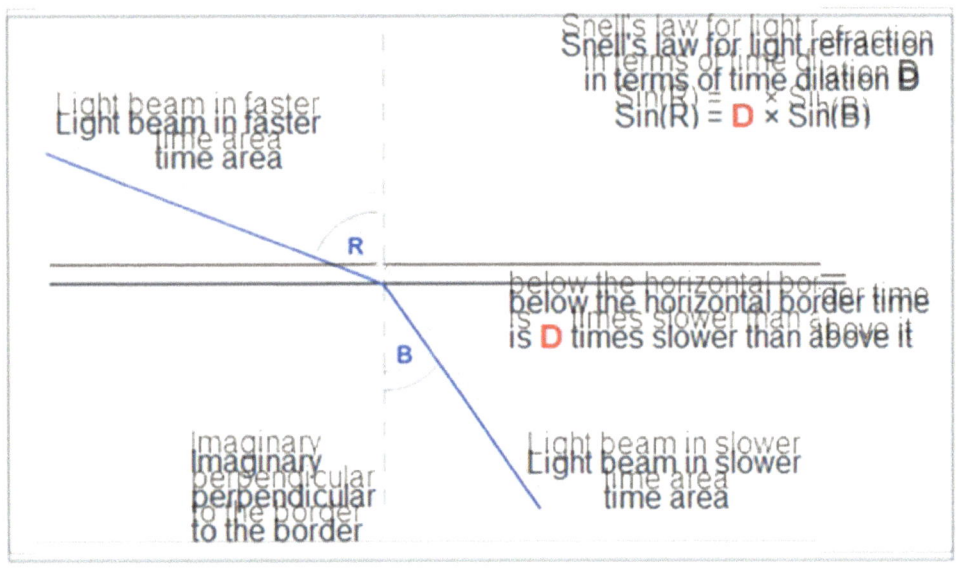

That can be derived geometrically, exploiting fact that a beam shoulders travel at different time speeds, covering different distances, and that difference is proportional to the time dilation factor B (distance between A2 and B2 is B times greater than distance between A1 and B1):

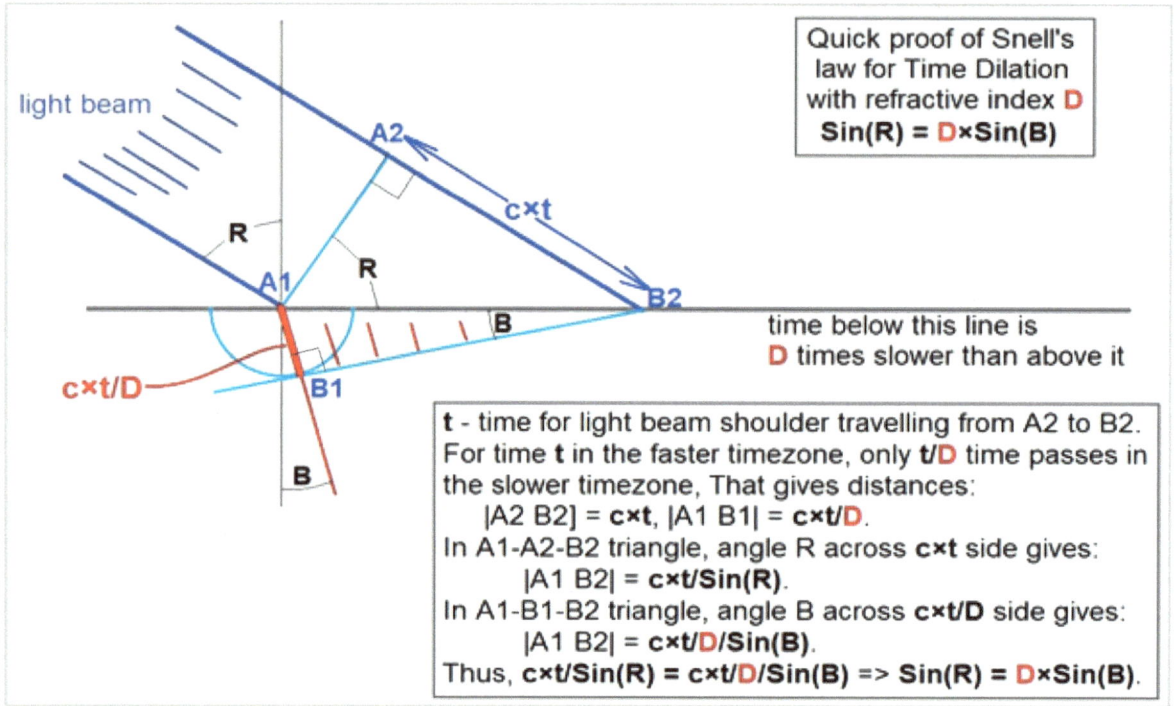

Quick proof of Snell's law for Time Dilation with refractive index **D**
Sin(R) = D×Sin(B)

time below this line is **D** times slower than above it

t - time for light beam shoulder travelling from A2 to B2. For time **t** in the faster timezone, only **t/D** time passes in the slower timezone, That gives distances:
|A2 B2] = **c×t**, |A1 B1| = **c×t/D**.
In A1-A2-B2 triangle, angle R across **c×t** side gives:
|A1 B2| = **c×t/Sin(R)**.
In A1-B1-B2 triangle, angle B across **c×t/D** side gives:
|A1 B2| = **c×t/D/Sin(B)**.
Thus, **c×t/Sin(R) = c×t/D/Sin(B) => Sin(R) = D×Sin(B)**.

P.S. The drawing above works for both cases: when the light beam goes from faster time to slower time, or from slower time to faster time.

P.P.S. Refraction is different both geometrically and physically from contraction or expansion of space speculated in relativity. For example, contraction/expansion below a borderline will break any line by changing **Tan** (but not **Sin**) of the incidence angle by the contraction/expansion factor:

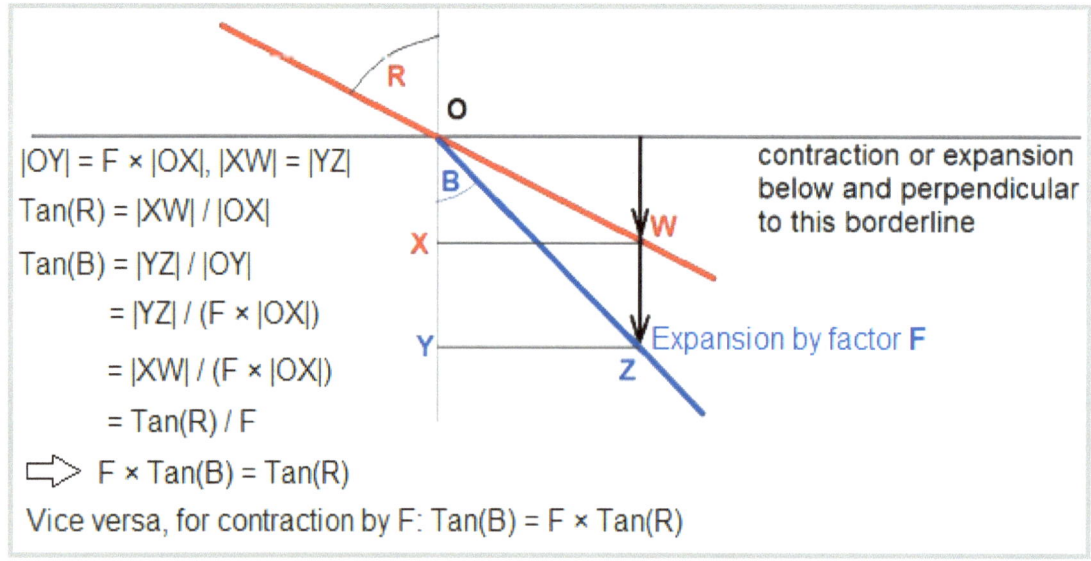

|OY| = F × |OX|, |XW| = |YZ|

Tan(R) = |XW| / |OX|

Tan(B) = |YZ| / |OY|

    = |YZ| / (F × |OX|)

    = |XW| / (F × |OX|)

    = Tan(R) / F

⟹ F × Tan(B) = Tan(R)

Vice versa, for contraction by F: Tan(B) = F × Tan(R)

contraction or expansion below and perpendicular to this borderline

Expansion by factor F

P.P.P.S. Refraction (and not Einstein's space curvature) is the real reason for some obscured-by-the-Sun stars being visible to astronomers during Solar eclipses.

2.3. Reflection

When an angle of incidence **B > arcsin (1/D)**, we have **D×Sin(B) > 1**, and such beam cannot cross a border from slower time to faster time, because otherwise we would have **Sin(R) > 1** for the angle of refraction **R,** and real **Sin** value cannot exceed **1**. Thus, such beam is reflected at the angle of reflection equal to the angle of incidence. Proof of that is no different from the classical proof because light beam stays in the same time:

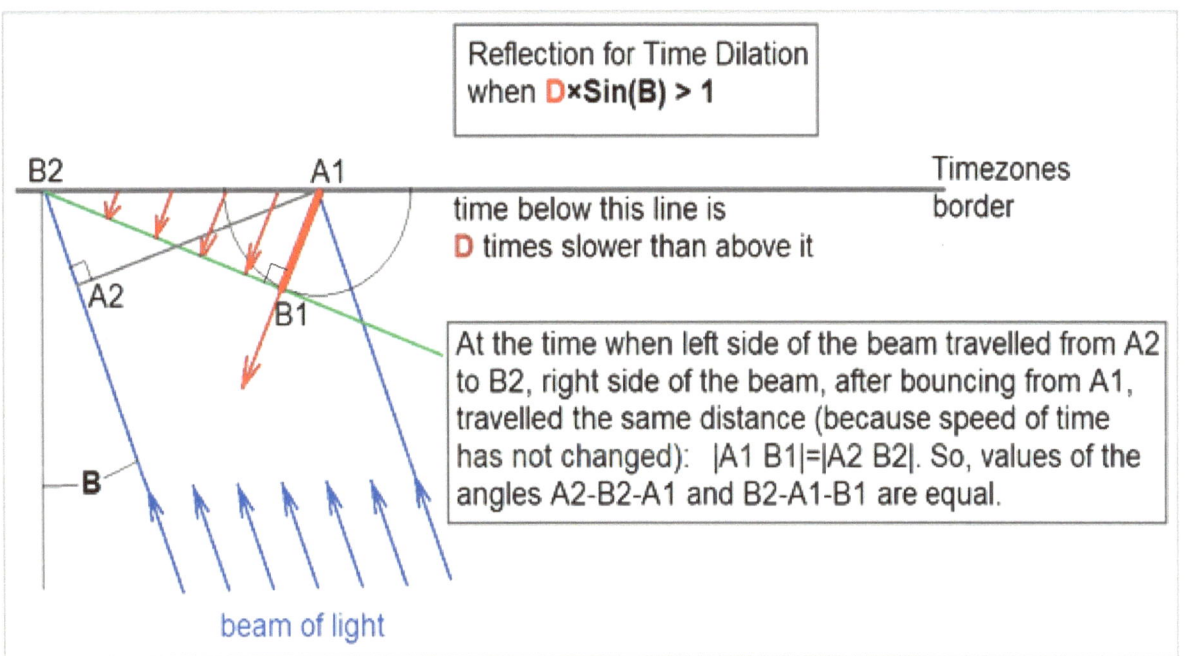

Reflection for Time Dilation when **D×Sin(B) > 1**

Timezones border

time below this line is **D** times slower than above it

At the time when left side of the beam travelled from A2 to B2, right side of the beam, after bouncing from A1, travelled the same distance (because speed of time has not changed): |A1 B1|=|A2 B2|. So, values of the angles A2-B2-A1 and B2-A1-B1 are equal.

beam of light

P.S. arcsin(1/D) separates incident angles at which refraction and reflection happens. It is the case for light coming from a slower time side, where **D>1,** and **arcsin(1/D)** exists. It is not the case for light coming from a faster time side, where **D<1** and **arcsin(1/D)** does not exist (**Sin** cannot be greater than **1**). Thus, light crossing from faster to slower time, at any angle of incidence, always refracts.

2.4. Multiple Images ★★

Let's return to the very first drawing in this book, but now let's consider not a narrow laser beam, but a widely shining light source (a star, for example, or a wide-angled flashlight):

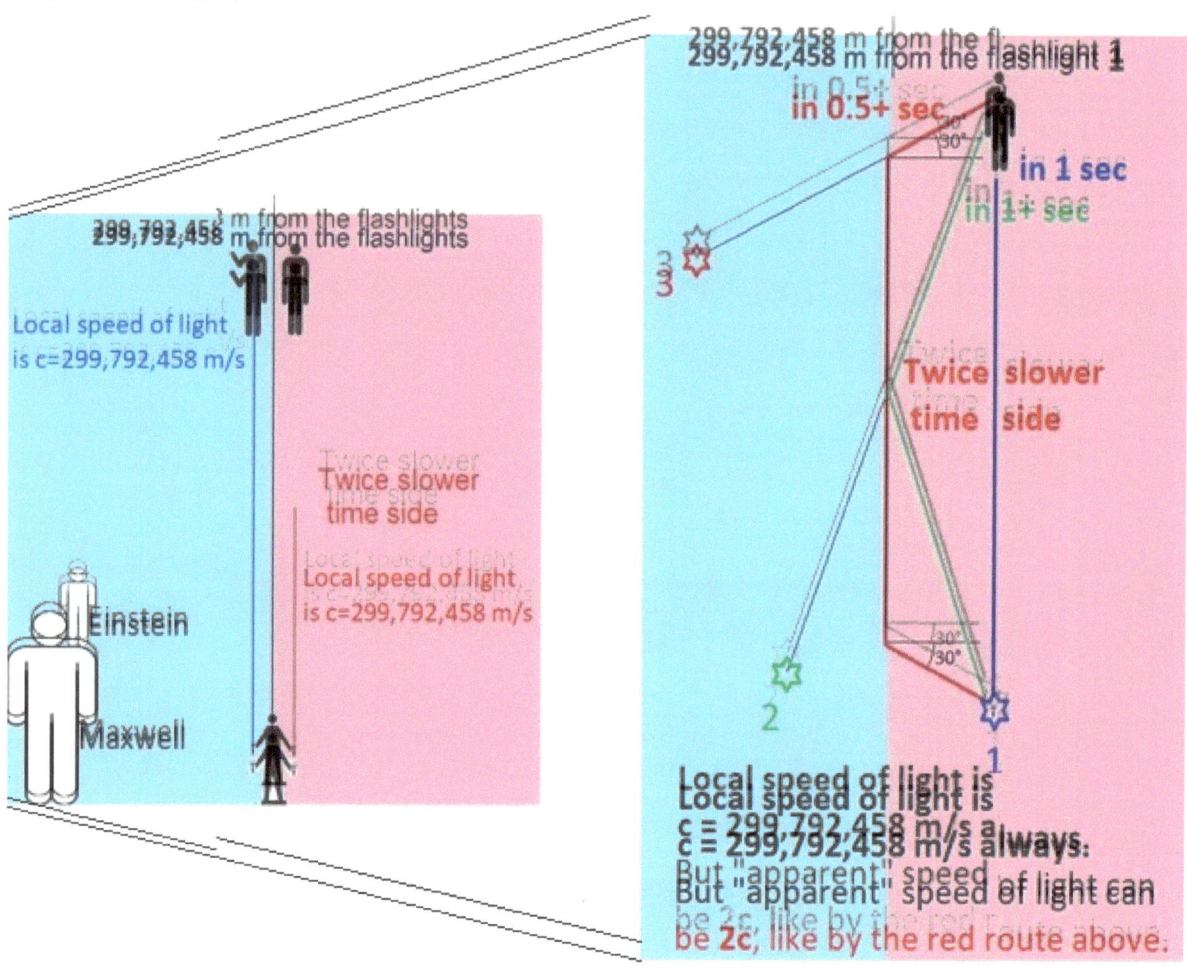

On the right-side picture, we have a closer look at what is happening on the slower time side:

- We consider only light paths that reach the observer.
- Light from the source **1** can reach the observer directly by blue trajectory in 1 sec (in local-to-the-pink-area time).
- Light from the source **1** is reflected from the border between the timezones as well, and it reaches the observer by green trajectory in a bit more than 1 sec. We assume that distance to the borderline is negligible in comparison with

299,792,458 m. Then the light source is visible somewhere at "position 2", besides the real position 1:

- Third, the red route is the most interesting, when light is twice refracted by Snell's law, with refractive index B ≡ 2, and **Sin(90°) ≡ 2×Sin(30°)**. Left shoulder of the red beam between refraction points travels in the twice-faster-time area. So, most of that path is covered in about a second of blue time, which is about half a second of pink time. Thus, this beam reaches the observer in a bit more than half a second (in local-to-the-pink area time units). Source of the light is visible at "position 3", in addition to positions 1 and 2:

- The observer thinks that 1, 2, and 3 are different light sources, when in fact there is only light source 1:

2.5. Visibility Angle

In 3.6 we will re-derive Einstein's formula for relativistic time dilation:

$$B \equiv 1/\sqrt{1 = v^2/c^2}$$

It states that time for a moving-at-speed-v object slows down by that factor **B**. Here we will use it to prove that a particle moving at velocity **v** can be seen only at **arccos(v/c)** angle:

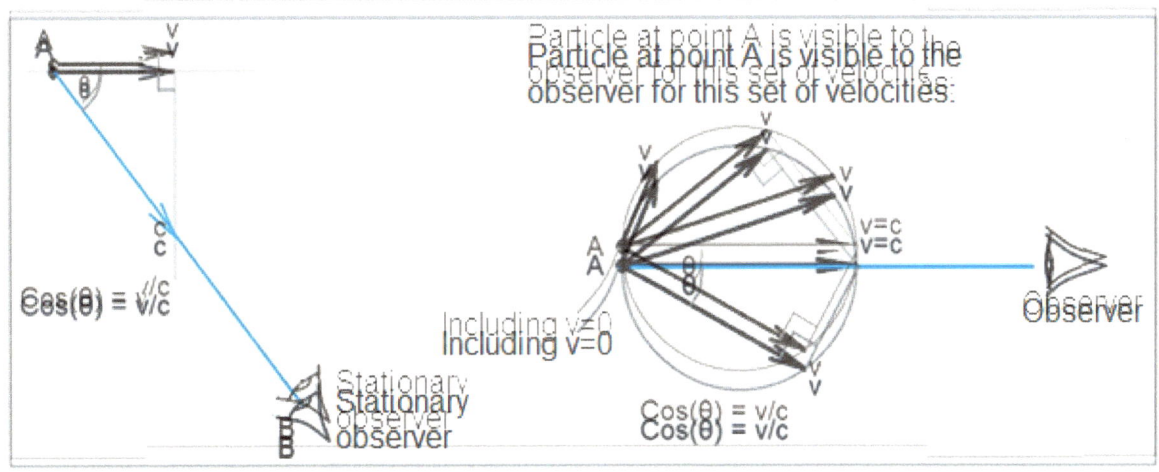

Drawing on the right shows that particle A

- Is always visible, when it is at rest;
- When its velocity **v** is small in comparison with the speed of light **c**, the particle is visible at an angle almost perpendicular to its velocity vector **v** (angle of visibility is a bit less than 90°);
- When this particle moves at a high speed close to **c**, line of sight is almost on the path of the particle (angle of visibility is close 0°).

Let's explore a particle that was at point **A** and had velocity **v** at that point (we don't care about its velocity before or after that). Let's say, we have two observers: a stationary observer at point **B** (see the drawing below), and another observer moving at constant velocity **v**: the second observer was at point **A** when the observed particle was there, and later he was at some point **E** at the moment when the first observer, stationed in **B**, noticed/saw the particle at the point **A** (after some time delay **t** for light to travel from **A** to **B**):

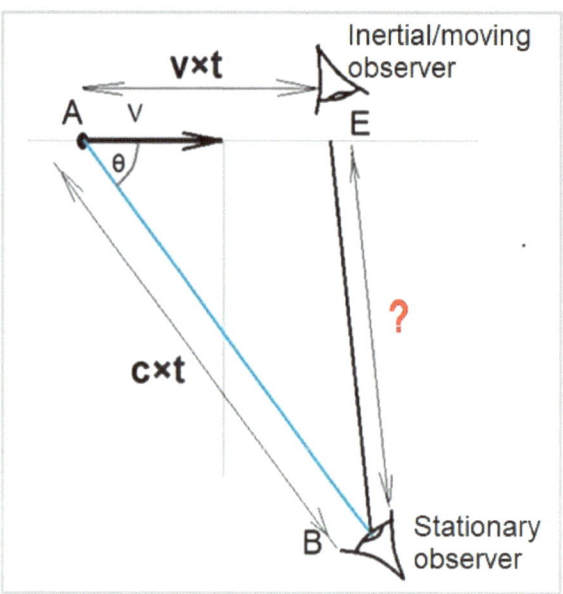

For the moving observer (between **A** and **E**), time slows down by a factor of **sqrt(1– v²/c²)**, where **sqrt** is square root, and in his frame of reference (where he thinks of himself as not moving) light has traveled not from **A** to **B**, but from **E** to **B**, and that took **t×sqrt(1–v²/c²)** time. Since we know all sides in the triangle **ΔABE**, we can find value **θ** of the angle **∠EAB**:

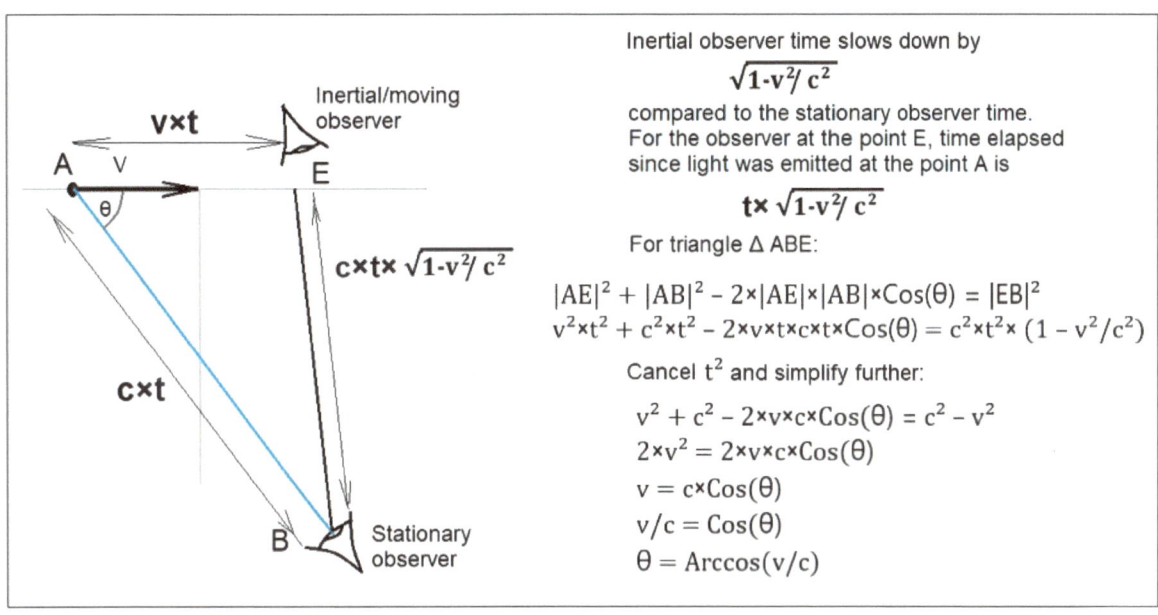

- $\theta = Arccos(v/c)$ **(2.5.1)**

P.S. In real life we do not experience such restrictions to visibility, why is that? Because we see light reflected from electrons whirling in atoms (and atoms themselves wobble, and electrons' velocities and thermal velocities of atoms are random). Thus, velocities of these electrons combined with the object's velocity are very random, therefore, combined velocities are much smaller than the speed of light **c**, with random directions (thus, for thousands of times a second, directions are close to perpendicular to an observer position), and at such combined velocities they are often visible. Electrons in atoms move at speeds of several hundred (up to a few thousands) km/sec. Thus, electrons' speeds are hundreds of times smaller than the speed of light **c**. The size of an atom is billions of times smaller than a meter, thus, we can see those whirling electrons billions of times per second, which is more than enough for our slow eyes to see the object all the time. No surprise: a century ago people watched movies at about 20 frames per second. But if an object itself moves at a speed of thousands km/sec, then speed of electrons in hundreds of km/sec cannot cancel such a high speed out, and such a fast moving object is visible at one angle, but invisible at another angle to its velocity. That solves the vanishing stars mystery: check Time Matters chapter 62.

P.P.S. To get physical explanation/intuition (besides geometrical) on why the angle of visibility is 90–° (less than 90°), check Time Matters chapter 99.

2.6. Concave Lensing ★

What happens with light emitted by distant galaxies a long time ago (when time in the Universe ran slower, as it was well explained in Time Matters), when our telescopes receive that light? As we discussed in 2.1, such light is redshifted. But besides that, refraction by Snell's law plays an interesting role here!

With no time dilation accounted for, light from a star or a galaxy S is received at points O and C at straight, not deflected, lines of sight (SO) and (SC):

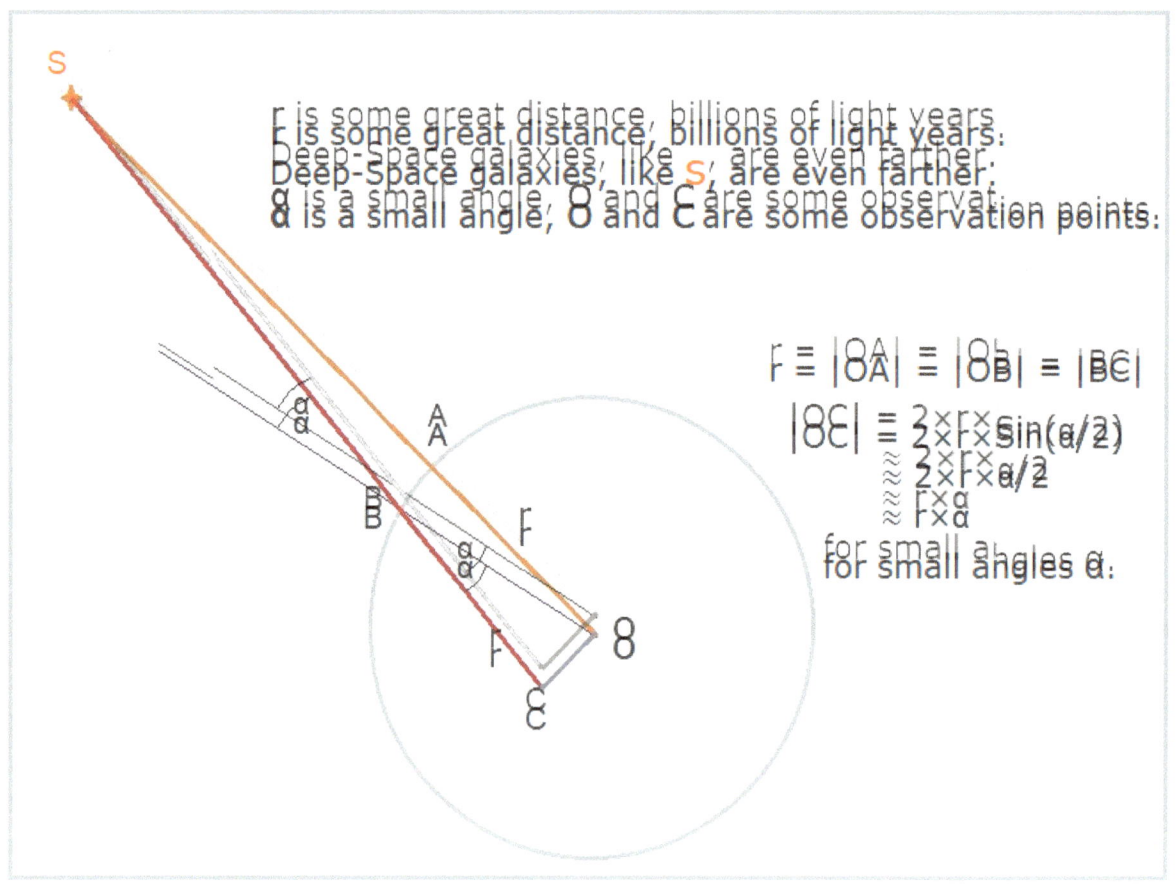

r is some great distance, billions of light years.
Deep-space galaxies, like s, are even farther.
α is a small angle, O and C are some observation points.

$$r \equiv |OA| \equiv |OB| = |BC|$$
$$|OC| \equiv 2 \times r \times \sin(\alpha/2)$$
$$\approx 2 \times r \times \alpha/2$$
$$\approx r \times \alpha$$

for small angles α.

Straight view picture

But in the past, time was slower than time nowadays, so starlight deflects by Snell's law, with time dilation factor β playing a role of refractive index, and with incidence angle α changing to refraction angle β > α:

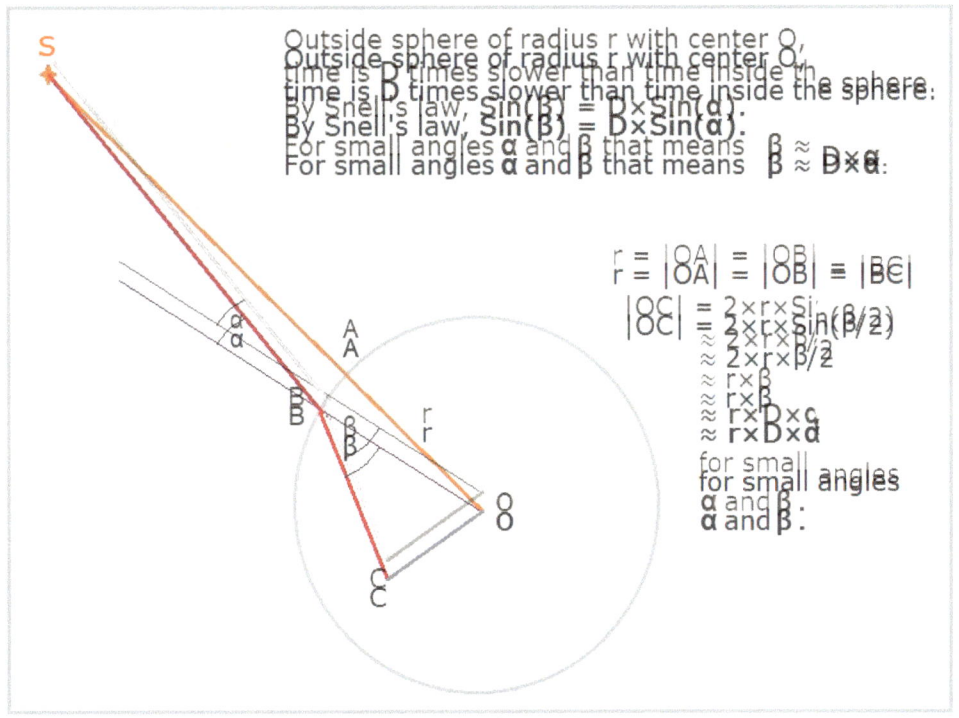

Outside sphere of radius r with center O, time is D times slower than time inside the sphere.
By Snell's law, $Sin(\beta) = D \times Sin(\alpha)$.
For small angles α and β that means $\beta \approx D \times \alpha$.

$r = |OA| = |OB| = |BC|$
$|OC| = 2 \times r \times Sin(\beta/2)$
$\approx 2 \times r \times \beta/2$
$\approx r \times \beta$
$\approx r \times D \times \alpha$
for small angles α and β.

Time dilated / refracted picture

In comparison to the no-time-dilation scenario, distance $|OC|$ has increased by factor D, and distance $|AB|$ has remained the same. Light source image appears at S1, which is closer than S, where the light was emitted from:

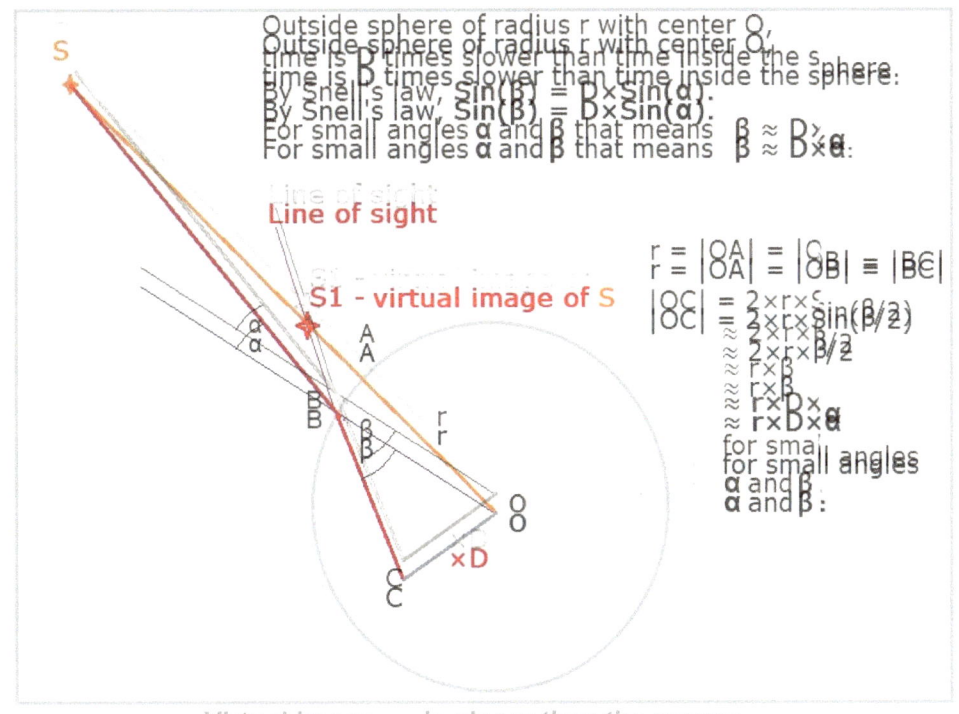

Outside sphere of radius r with center O, time is D times slower than time inside the sphere.
By Snell's law, $Sin(\beta) = D \times Sin(\alpha)$.
For small angles α and β that means $\beta \approx D \times \alpha$.

Line of sight

S1 - virtual image of S

$r = |OA| = |OB| = |BC|$
$|OC| = 2 \times r \times Sin(\beta/2)$
$\approx 2 \times r \times \beta/2$
$\approx r \times \beta$
$\approx r \times D \times \alpha$
for small angles α and β.

×D

Virtual image **S1** is closer than the source **S**

Image of the galaxy appears smaller than its actual size. In optics, such effect is called "concave lensing":

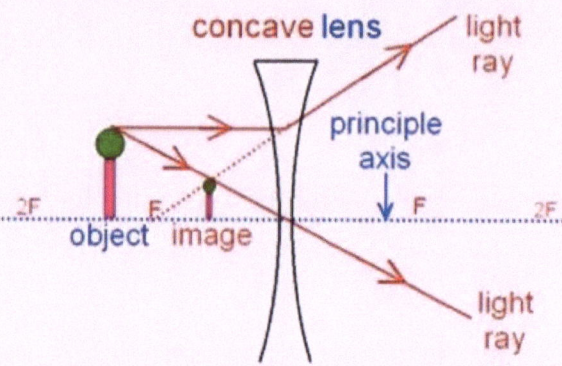

A concave lens is a diverging lens which
makes the rays of light disperse and spread further apart.
It does the opposite of a convex lens.

You can see that the image is not the same as the object.
The image is called virtual because the light rays
never really cross there (compare this with a real image).
The image is called upright because it is the right way up
(compare this with an inverted image). The image is smaller
than the object and on the same side of the lens as the object.

Concave lensing: closer and smaller image https://www.gcsescience.com/pwav29.htm

3. TIME POTENTIAL

Physics is mathematical not because we know so much about the physical world, but because we know so little.

Bertrand Russell

3.1. Potential = 0.5×c²/D²

In chapter 1, we got formula (1.3) **c/D** for apparent speed of light, or light wave. In classical mechanics, there is Newton–Laplace formula for a wave speed in any medium:

$$c = \text{sqrt}(\gamma \times P/\rho).$$

"**c**" denotes here speed of wave, **P** – pressure in the medium, **ρ** – density of the medium, and **γ** is a numerical factor, which we will discuss at the end of this section. By replacing the speed of wave "**c**" with speed of light wave **c/D**, we have:

$$c/\mathbf{D} = \text{sqrt}(\gamma \times P/\rho).$$

Now, remove **sqrt** by squaring both sides:

$$c^2/\mathbf{D}^2 = \gamma \times P/\rho.$$

At point X:

$$c^2/\mathbf{D}(X)^2 = \gamma \times P(X)/\rho.$$

At point X+Δ:

$$c^2/\mathbf{D}(X+\Delta)^2 = \gamma \times P(X+\Delta)/\rho.$$

Subtract those:

$$c^2/\mathbf{D}(X+\Delta)^2 - c^2/\mathbf{D}(X)^2 = \gamma \times (P(X+\Delta) - P(X))/\rho.$$

Divide both sides by Δ, and multiply and divide the right side by 1m²:

- **(c²/D(X+Δ)² − c²/D(X)²) / Δ = γ×(P(X+Δ) − P(X)) ×1m² / (ρ×1m²×Δ)**   **(3.1.1)**

(P(X+Δ)–P(X))×1m² = –F(X), where F is the resulting force acting on 1m²×Δ layer.

Negative sign before F(X) indicates opposite-to-X direction of F, when P(X+Δ) > P(X):

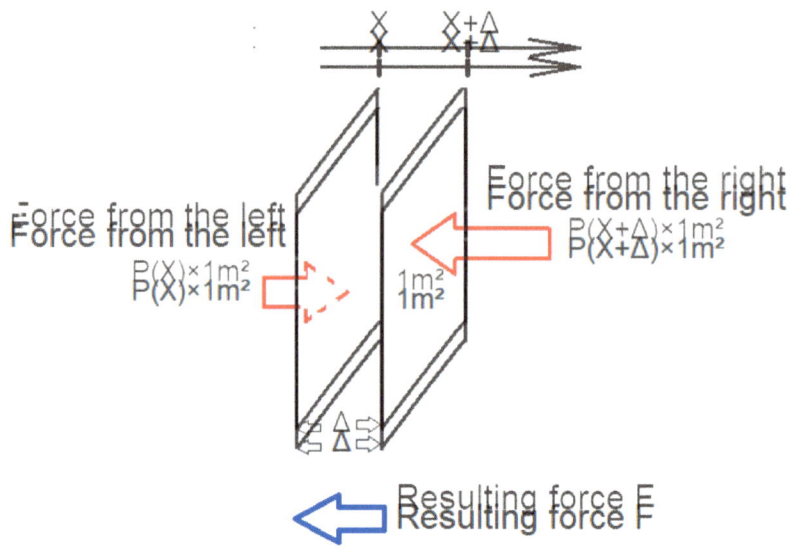

When $P(X+\Delta) < P(X)$, $P(X+\Delta) - P(X)$ is negative, $-F(X)$ is negative too; therefore, $F(X)$ is positive, thus, resulting force F points in the same direction as X.

$\rho \times 1m^2 \times \Delta$ represents the mass of this layer. Force divided by mass is acceleration or derivative of potential with negative sign; thus:

$$\gamma \times (P(X+\Delta) - P(X)) \times 1m^2 / (\rho \times 1m^2 \times \Delta) \equiv \gamma \times Potential'(X)$$

Apostrophe ' denotes derivative by location (by X in this case).

And since $(c^2/\rho(X+\Delta)^2 - c^2/\rho(X)^2) / \Delta \equiv (c^2/\rho(X)^2)'$, formula (3.1.1) simplifies to:

$$(c^2/\rho(X)^2)' \equiv \gamma \times Potential'(X)$$

Getting rid of derivatives on both sides:

$$c^2/\rho^2 \equiv \gamma \times Potential$$

(plus a constant, but in physics, + constant is ignored for a potential).

Now, to the factor γ: it is linked to the degrees of freedom f of the medium "particles", as $\gamma \equiv 1+2/f$ (see details in Wikipedia: *Relation with degrees of freedom*). For a particle moving at the speed of light, there are only two degrees of freedom: it can oscillate only in the plane perpendicular to the velocity of the particle, because oscillation in the direction of motion will change the speed of the particle, and this speed (which should be the speed of light) cannot change. Check more formal explanation in

Quote from a theoretical physicist:

> *"In summary, massless particles have only two degrees of freedom because their spin states are representations of the two-dimensional rotation group SO(2), which is Abelian..."*

Thus, with $f \equiv 2$ and $\gamma \equiv 1+2/f \equiv 2$, we have:

$$c^2/B^2 \equiv \gamma \times \text{Potential} \equiv 2 \times \text{Potential}$$

- **Potential** $\equiv 0.5 \times c^2/B^2$:

3.2. Gravity $g = -(0.5c^2/D^2)' \approx c^2 \times D' \sim \nabla D$

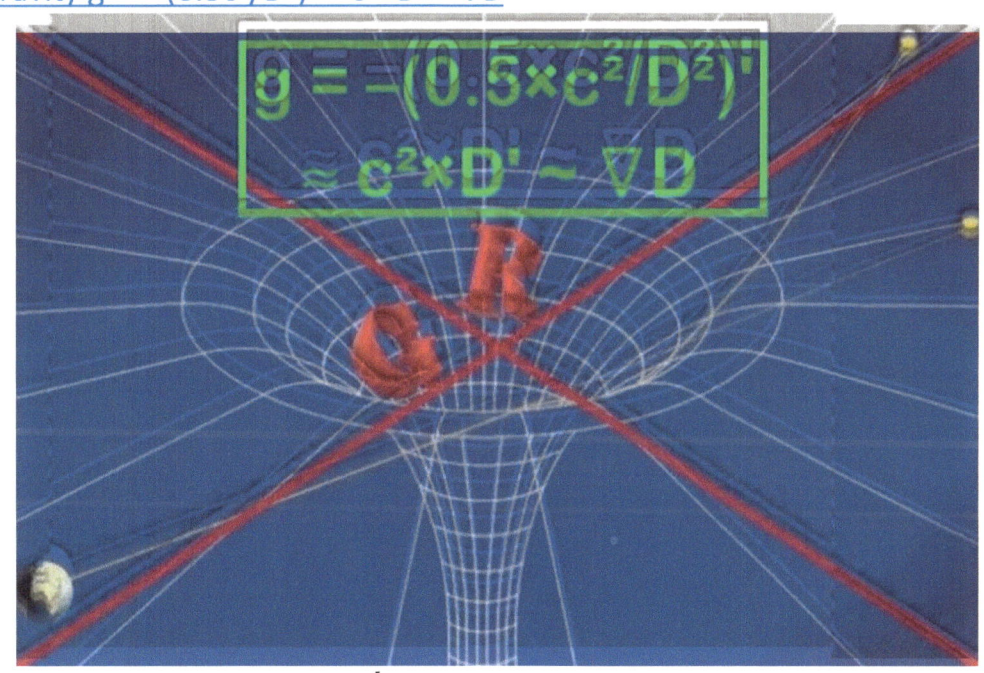

$$g \equiv -(0.5 \times c^2/D^2)'$$
$$\approx c^2 \times D' \sim \nabla D$$

Gravity g depends only on B' or ∇B (derivative by location or gradient of B):

By classical mechanics, derivative by location of **–Potential** is acceleration. So, derivative by location of the time potential gives acceleration g:

$$g \equiv -(0.5c^2 / B^2)' \equiv c^2 \times B' / B^3$$

When time dilation $B \approx 1$ (for example, near the Sun we have $B \equiv 1.000002$):

$$g \equiv -(0.5c^2 / B^2)' \equiv c^2 \times B' / B^3 \approx c^2 \times B' \approx 9 \times 10^{16} \times B' \ \text{m}^2/\text{sec}^2 \approx \nabla B$$

"~" denotes proportionality: in the line above, **g** is proportional to **∇D** (or **D'**) with **9×10¹⁶** coefficient. We know that **g ≈ 9.8** m/sec² on the Earth, thus, near the Earth:

g ≈ 9.8 m/sec² ≈ **9×10¹⁶×D'** m²/sec²   => **D'×1m ≈ 10⁻¹⁶**.

We just explained the book cover images.

3.3. Gravitational Time Dilation D = exp(G×M/(R×c²))

Newton's law for gravitational acceleration **g** at distance **R** from mass **M**

g = – G×M/R²

does not account for time dilation **D**. Formula (1.2) from chapter 1 says that apparent acceleration is local acceleration divided by **D²**, so observed acceleration is:

- **g = – G×M/R²/D² = – G×M/(R²×D²).**   (3.3.1)

The previous section started with another formula for **g**:

g = – (0.5×c²/D²)'.

Thus, **g = – (0.5×c²/D²)' = – G×M/(R²×D²)**

0.5×c²×(D⁻²)' = G×M/(R²×D²)

0.5×c²×(–2D'×D⁻³) = G×M/(R²×D²)

– c²×(D'/D) = G×M/R²

– c²×(ln(D))' = G×M×(–1/R)'

c²×ln(D) = G×M×(1/R), no +Constant, since both sides are 0 for R=∞ and D=1

ln(D) = G×M/(R×c²)

- **D = exp(G×M/(R×c²))**, where exp(x) = eˣ.   (3.3.2)

From (3.3.1) and (3.3.2) we get Dr. Robinson's gravitational formula:

- **g = – G×M/(R²×exp(2GM/(Rc²))).**   (3.3.3)

Now, finding time dilation **D** near the Sun becomes an easy exercise, just by replacing letters in formula (3.3.2): G = 6.674×10⁻¹¹, M= 2×10³⁰, R= 7×10⁸, c=3×10⁸,

$$\mathbf{D} \approx \exp(6.674×10^{-11}×2×10^{30}/(7×10^{8}×3^{2}×10^{16})) \approx \exp(2.12×10^{-6}) \approx 1.000002.$$

And you can calculate the time dilation near the Earth for M=5.97×10²⁴ kg and R=6.37×10⁶ m, using Wolfram calculator:

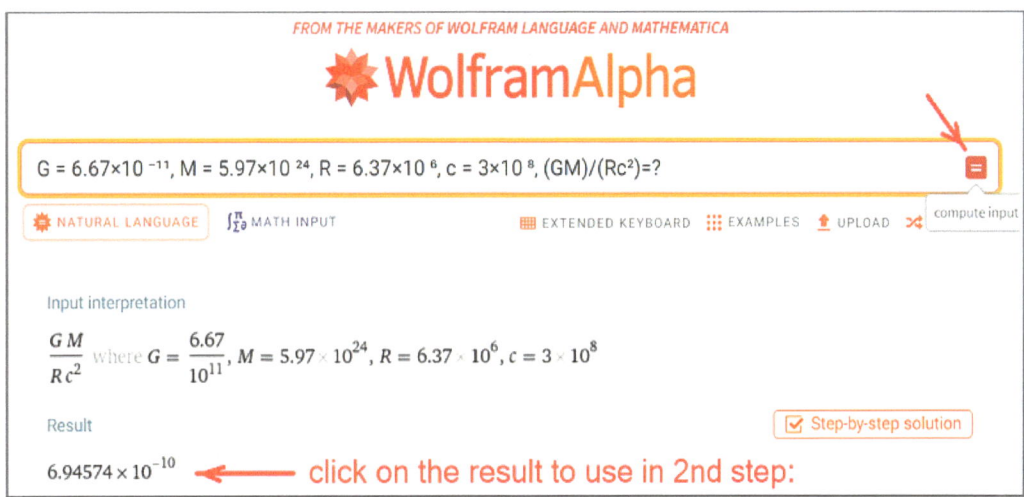

In the Wolfram calculator, you can use ^ for power: 2^10 = 2¹⁰. And you can chain calculations.

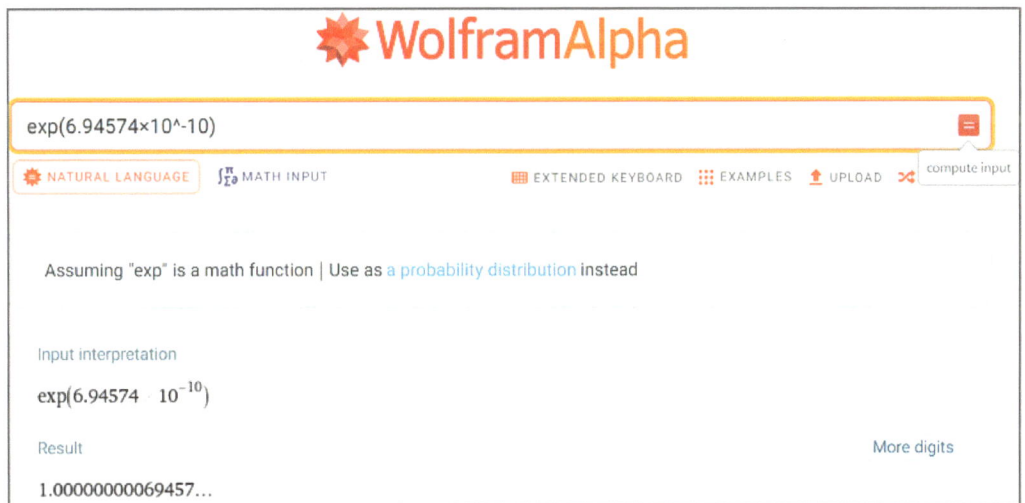

P.S. Formula **D = exp(G×M/(R×c²))** debunks "theoretical Black Holes", because it shows that there is no so-called "Event Horizon" where time stops (and time dilation **D** is infinite). Time can slow down by a finite factor **D**, but it does stop: **D ≠ ∞** !

P.P.S. Formula $B \equiv \exp(G\times M/(R\times c^2))$ with a constant mass M works outside that mass, when $R \geq R_0$, where R_0 is the radius of this mass M. By [Newton's shell theorem], mass m at distance $R < R_0$ from the center of the body-mass M is attracted only by mass of the ball of radius R (not by the whole ball with radius R_0), which has mass less than M. Volume of the ball with radius R is proportional to R^3. Assuming that the density at the center of the balls is finite, we have mass of ball with radius R is proportional to

$$\approx M\times(4/3\times\pi\times R^3) / (4/3\times\pi\times R_0^3) \equiv M\times R^3 / R_0^3,$$

where M and R_0 are constants. Thus, formula for time dilation B inside the object at distance $R < R_0$ from its center is:

$$B \approx \exp(G\times(M\times R^3 / R_0^3)/(R\times c^2)) \equiv \exp(G\times M\times R^2/R_0^3/c^2)) < \exp(G\times M/R_0/c^2)).$$

Even inside a celestial body, time dilation B is finite (because $\exp(G\times M/R_0/c^2))$ is finite and there is no division by R with R near 0).

P.P.P.S. Newton's gravity $g \equiv -G\times M/R^2$ keeps planets in elliptical orbits, but with $g \equiv -G\times M/(R^2\times B^2(R))$ and force F (by Newton's second law):

$$F \equiv -G\times m\times M/(R^2\times B^2(R)), \text{ where value of } B(R) \equiv \exp(G\times M/(R\times c^2)), \quad (3.3.4)$$

$B(R)$ increases when the distance R decreases (or, vice versa, $B(R)$ decreases when R increases), elliptical orbits undergo shift/precession:

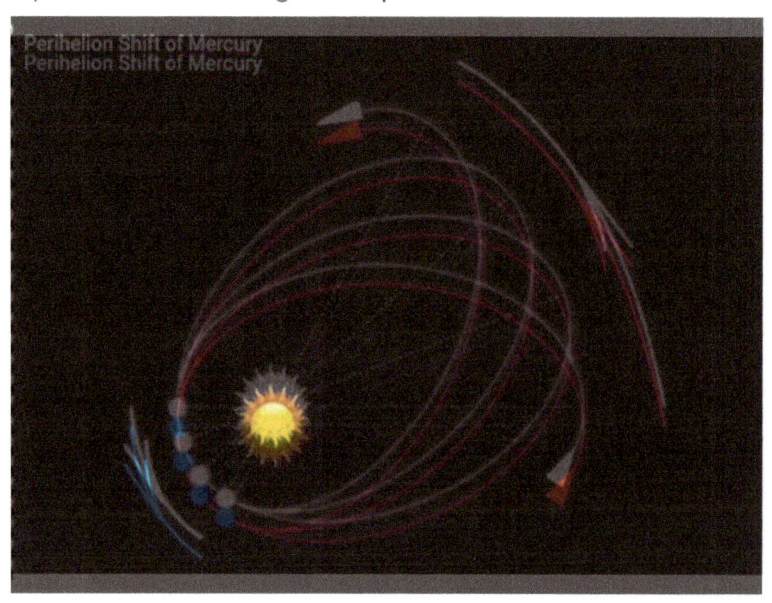

- When the planet is moving towards the Sun, with increasing **D(R)** (in denominator of the force in (3.3.4)), gravity is gained at a lower than by classical formula rate, that is not enough to retain a planet in its classical orbit. That causes the planet to shift outside the anticipated classical orbit.

- When the planet is moving away from the Sun, with increasing **R** and decreasing **D(R)** (in denominator of the force in (3.3.4)), gravity weakens slower than classically expected, so the planet is pulled inside the classical ellipse by larger than expected force.

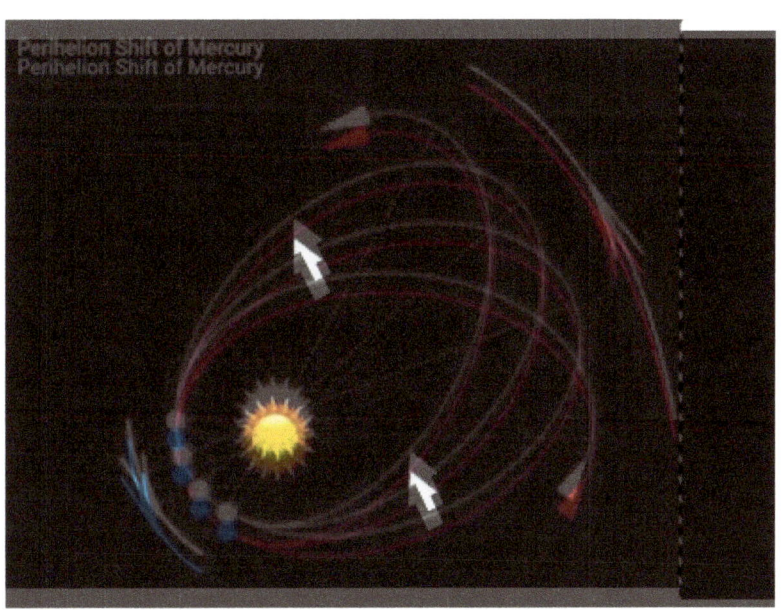

It is good exercise for a Computer Science or Physics freshman/sophomore to write an animation in Python for Mercury's precession, using only the formula

$$g = -G \times M/(R^2 \times \exp(2GM/(Rc^2))).$$

3.4. Escape Velocity = c × sqrt(1−exp(−2G×M/(R×c²))) < c

Escape velocity, derived in classical and relativistic mechanics, **v = sqrt (2GM/R)**, where **sqrt** is square root, **G** is the gravitational constant, **M** is the mass of the attractor, **R** is the distance to the attractor, has a known problem: for massive and dense attractors (like infamous "Black Holes"), 2GM/R value can be greater than c², and then **v ≥ c**. Such **v** is impossible to achieve, because c is local speed limit. Therefore, the mainstream claims that some (massive) bodies at some (close) distances are inescapable. Is it for real, or is it just a math mistake? No worries – it is just a math mistake. Let's start with a classical derivation of escape velocity and then correct it. Let's calculate the escape work done against Newton's gravitational force GMm/x^2, where x is distance to the attractor and m is the mass of the escaping body, when that escaping body moves from distance R to infinity:

$$\int_R^\infty GMm/x^2\,dx = GMm\int_R^\infty x^{-2}\,dx = -GMm{\times}x^{-1}\Big|_R^\infty = GMm/R$$

By the energy conservation principle, this work should be equal to the change of kinetic energy $mv^2/2$ of the escaping body (where v is its velocity when it was at the distance R to the attractor, and with its final velocity close to 0, when it is far away, at ∞), thus:

$$GMm/R = mv^2/2 \implies v = \mathrm{sqrt}(2GM/R)$$

This completes the classic derivation.

The correction lies in using Dr. Robinson's precise formula for gravitational force:

(3.3.4): $GMm / (R^2{\times}\exp(2GM/(Rc^2)))$

instead of Newton's gravitational force GMm/R^2 in calculation of the escape work:

$$\int_R^\infty GMm/(x^2{\times}\exp(2GM/(xc^2)))\,dx = GMm\int_R^\infty x^{-2}{\times}\exp(-2GM/c^2 \times x^{-1})\,dx =$$

$$= -GMm\int_R^\infty \exp(-2GM/c^2 \times x^{-1})\,d(x^{-1}) = -GMm\int_{1/R}^0 \exp(-2GM/c^2 \times y)\,dy =$$

$$= GMm \times c^2/(2GM) \times \exp(-2GM/c^2 \times y)\Big|_{1/R}^0 = mc^2/2 \times (1-\exp(-2GM/(Rc^2)))$$

Thus, $mc^2/2 \times (1-\exp(-2GM/(Rc^2))) = mv^2/2$. Therefore,

- Escape velocity $v = c \times \mathrm{sqrt}(1-\exp(-2GM/(Rc^2))) < c$ $\qquad$ (3.4.1)

The classical formula $v = \mathrm{sqrt}(2GM/R)$ for escape velocity is just an approximation for formula (3.4.1), by Taylor series approximation $\exp(x) \approx 1+x + \dots$ for small x:

$$v = c \times \mathrm{sqrt}(1-\exp(-2GM/(Rc^2))) \approx c \times \mathrm{sqrt}(1 - (1-2GM/(Rc^2)+\dots))$$

$\approx$ c × sqrt (2GM / (Rc²)) = sqrt (2GM/R).

Such approximation is appropriate only for small values x, i.e. small GM /(Rc²).

3.5. Action = − D² × Reaction

Newton's inverse square law for gravitational force **F** between masses **m** and **M** separated by distance **R** is:

- **F = − G×m×M/R²**   **(3.5.1)**

(which is the same law for gravitational acceleration **g = − G×M/R²**, because **F = g×m** by Newton's second law for motion). With (3.3.1) correction, we have:

- **F = − G×m×M/(R²×D²),**   **(3.5.2)**

where **D** is time dilation near the object to which force **F** is applied (and acceleration **g** of the object observed). Formula (3.5.1) is perfectly symmetrical, for example, the Sun attracts the Earth by the force of the same value as the Earth attracts the Sun (and forces point to each other). Such symmetry is known as Newton's third law of motion:

Action = − Reaction.

But symmetry goes away with formula (3.5.2), because of different **D** values, for example, near the Sun D $\approx$ 1.000002, but near the Earth D $\approx$ 1 (or D $\approx$ 1.0000000007 if an observer is outside the Earth). Thus, force by which the Earth attracts the Sun is

F = − G×m×M/(R²×1.000002²),

which is less than the force by which the Sun attracts the Earth:

F = − G×m×M/R².

The same is true for attraction between the Sun and any other planet, Mars for example (near Mars, **D<1** from the Earth's perspective).

Let's modify Newton's third law to account for time dilation, and not only for gravitational forces, but for whatever remote (thus, possibly in different timezones) action and reaction (for example, Coulomb electrostatic force). Our modified Newton's third law is:

- $F1 \times B1^2 \equiv - F2 \times B2^2,$ (3.5.3)

where **F1** is action/force acting on **object 1**; **B1** is time dilation near **object 1**; **F2** is reaction/force acting on **object 2**; **B2** is time dilation near **object 2**. From an observer on object 1 perspective **B1 ≡ 1**; thus (3.5.3) is simplified to:

- **Action** $\equiv - B^2 \times$ **Reaction**   (3.5.4)

with **B** as **B2**. (3.5.3) and (3.5.4) are true for gravity, because $F1 \times B1^2$ and $F2 \times B2^2$ are both equal to $G \times m \times M/R^2$; with B^2 in the denominator of formula (3.5.2) canceled with B^2 factor in (3.5.3 – 3.5.4). And for other types of forces, a similar cancellation of B^2 also works when applying the formula (1.2) apparent acceleration ≡ local acceleration / B^2 for forces.

P.S. You might object that formula (3.5.2), from local perspective of **B ≡ 1**, is always the same as (3.5.1) for that observer: for on-the-Sun observer, the Sun is attracted by the same force $G \times m \times M/R^2$ as the force by which the Earth is attracted to the Sun for on-the-Earth observer. Then Newton's **Action ≡ −Reaction** still holds. Mistake with such a line of reasoning is: local-to-the-Earth and local-to-the-Sun units of measure, which depend on local second, (like, for example, the unit of acceleration m/sec²), are actually different units. And values in different units cannot be compared unless they are converted to the same unit.

P.P.S. The same applies to the energy unit <mark>Joule</mark> measured in kg×m²/<mark>sec</mark>²: it <mark>is not a universal / absolute</mark> unit, because it is in local time units. And physical constants expressed in time-dependent units (for example, <mark>Planck constant</mark> in kg×m²/<mark>sec</mark>) are <mark>not universal / absolute</mark> constants.

3.6. Relativistic Time Dilation $D = 1 / \sqrt{1 - v^2/c^2}$

Having apparent speed of light c/B and no length contraction, we have to review, re-derive, or alter a well-known formula for relativistic time dilation:

$$B \equiv 1/\sqrt{1 = v^2/c^2}$$

In 3.1 we derived formula for time/gravitational Potential $\equiv 0.5 \times c^2/B^2$, which solved physics of gravity without curving space:

$$g \equiv -(0.5 \times c^2/B^2)' \equiv -\nabla(0.5 \times c^2/B^2).$$

Symbol $'$ and symbol ∇ both denote derivative by location. Let's consider an object that starts at zero speed and uniformly (at a constant acceleration g) accelerates to a speed v in t seconds. Acceleration, by definition, is change in speed per second:

$$g \equiv v/t.$$

Using [the formula for uniform acceleration](#)

$$S(t) \equiv S_0 + v_0 t + g t^2/2,$$

for initial $S_0 \equiv 0$; $v_0 \equiv 0$; for the distance $S(t)$ (let's denote it as L) covered by this object, we have:

$$L \equiv g \times t^2/2 \equiv (v/t) \times t^2/2 \equiv v \times t/2.$$

Integral of g by location (in this case by variable x from 0 to L) is:

$$\int_0^L g \, dx \equiv L \times g \equiv (v \times t/2) \times (v/t) \equiv v^2/2$$

(the integral above is equal to $L \times g$ because g is the constant acceleration there). And since $g \equiv -(0.5 \times c^2/B^2)'$, this integral is the same as

$$\int_0^L -(0.5 \times c^2/B^2(x))' \, dx \equiv 0.5 \times c^2/B^2(0) - 0.5 \times c^2/B^2(L)$$

Initial time dilation $B(0) \equiv 1$. Let's find the final value $B(L)$, denoting it as B (we are not interested in intermediate values of $B(x)$):

$$\int_0^L - (0.5 \times c^2 / D^2(x))' \, dx = 0.5 \times c^2 - 0.5 \times c^2 / D^2$$

Since all integrals above are the same, we have:

$$v^2/2 = 0.5 \times c^2 - 0.5 \times c^2 / D^2 \Rightarrow v^2/c^2 = 1 - 1/D^2 \Rightarrow 1/D^2 = 1 - v^2/c^2$$

$$D^2 = 1/(1 - v^2/c^2) \Rightarrow D = 1/\sqrt{1 - v^2/c^2}$$

We got the same formula for time dilation **D** as Einstein's Lorentz factor

$$\gamma = 1/\sqrt{1 - v^2/c^2}$$

But here we do not need length contraction (by Lorentz factor or any other factor).

P.S. Time dilation and **local speed of light constant c** stay, but not length and space metric manipulations. The only spatial effects, valid in this context, are optical effects discussed in chapter 2, which have nothing to do with spatial contraction or expansion as noted in the P.P.S to 2.2.

P.P.S. With acceleration g > 0, **D** value increases; with deceleration g < 0, **D** value decreases; with g = 0 (without acceleration or deceleration), **D** value does not change. That is enough to explain Twin Paradox, since acceleration and deceleration are needed for twins to meet and compare their clocks eventually.

4. REVISED MASS AND ENERGY FORMULAS

4.1. Mass $m = m_0 \times D^2$

$$m = \frac{m_0}{\sqrt{1 - v^2/c^2}} \Rightarrow E = mc^2$$

$$m = \frac{m_0}{1 - v^2/c^2} \Rightarrow E = mc^2/2$$

You might have heard of formula $m = m_0/\text{sqrt}(1-v^2/c^2)$, showing that mass m of an accelerated object increases to infinity when the object's velocity v approaches the speed of light c (because $1-v^2/c^2$ in the denominator comes close to zero, division by which gives infinity). Here m_0 denotes the rest mass or initial mass of the object when it does not move, sqrt denotes square root. Where does this denominator come from?

$$D = 1/\sqrt{1 - v^2/c^2}$$

That is rate of time dilation (that we have just confirmed in 3.6): time in fast moving objects slows down by factor D. That was Einstein's first discovery.

https://www.youtube.com/watch?v=ZyYqyYAKGC0

Einstein's formula for mass is basically $m = m_0 \times B$. But there is a mistake and the formula shall be corrected to $m = m_0 \times B^2$! Why is that?

Mass measures inertia, which is an object's response / opposition to a force applied. This response is observed through the change in the object velocity, which is called acceleration. Newton's 2nd law of motion $F = m \times a$ is about force F causing change in velocity of the object, which is acceleration a. Heavier objects react less than lighter objects when the same force is applied to them. If mass increases by a factor x, then acceleration decreases by the same factor – that is the meaning of Newton's 2nd law for a "fixed amount of force": we can say that mass and acceleration are inversely related. Now, what happens when time slows down by a factor B? We already figured that out in the first chapter:

(1.2): Apparent a = Local a / B^2.

Acceleration, observed from a faster time perspective (faster than local to the object time), is B^2 times smaller than the local acceleration. Thus, mass as measurement of inertia (resistance to force) increases by factor B^2 (as mass m is inversely related to acceleration a). We just proved $m = m_0 \times B^2$, or

- Apparent mass m = Local mass $m_0 \times B^2$ (4.1.1)

4.2. $E = m \times c^2 / 2$

Recently I checked the original and other derivations of $E = mc^2$ formula, and all of them have time-related mistakes. Let's start with the original proof.

Einstein derived $E = mc^2$ formula using a thought experiment (you can check https://www.youtube.com/watch?v=hW7DW9NIO9M for a popular 2-minute video and https://www.youtube.com/watch?v=vopowgznsXw for a complete presentation of his proof). Einstein compares the kinetic energy of an object after it radiates at least two photons (in opposite directions, so that does not change the speed of the object). And he calculates energy in two ways:

Chapter 4: REVISED MASS AND ENERGY FORMULAS

1) From a moving-at-speed-v-observer point of view: the initial object kinetic energy KE_1 minus energy E of the radiated photons multiplied by $1+v^2/(2c^2)$. Factor $1+v^2/(2c^2)$ is attributed to Doppler effect (check the [video](#) and screenshot below);

2) From a static observer perspective, Einstein accounts energy $=E$ of emitted photons first, then from the moving observer perspective, Einstein adds kinetic energy KE_2 of the object, which became a bit lighter after photons emitted (so $KE_2 < KE_1$):

$$KE_1 = E \times (1+v^2/(2c^2)) = -E + KE_2$$

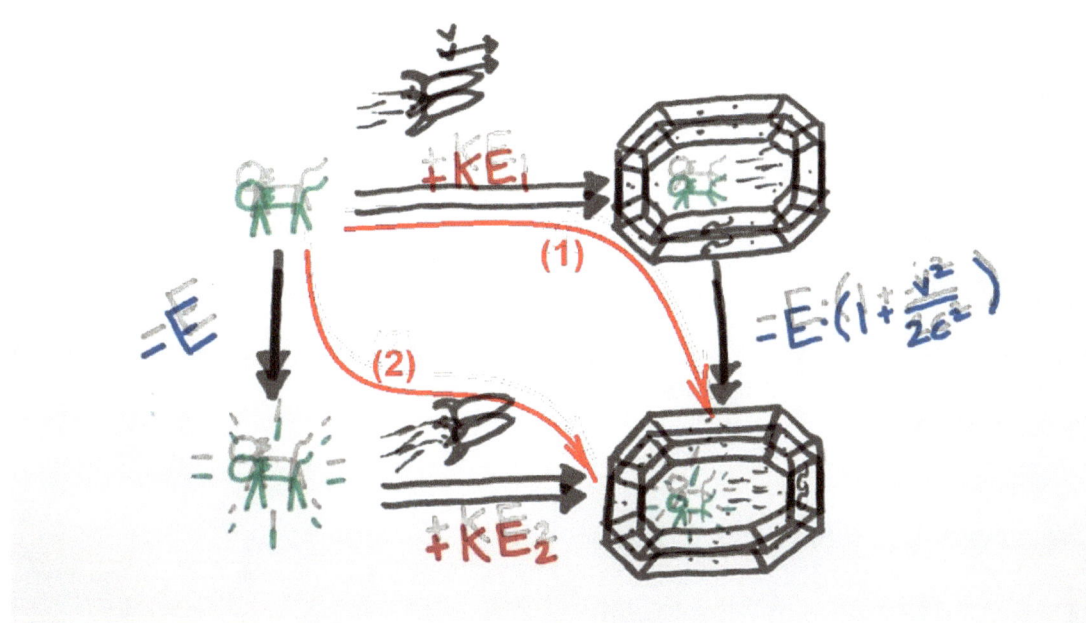

Watch 2-min Einstein's Proof of E=mc² at
https://www.youtube.com/watch?v=hW7DW9NIO9M

Here is the mistake: in 2), energy of photons denoted as $-E$ is in units of energy related to static time, but energy of the object denoted as $+KE_2$ is in units of energy related to dilated/slower time in the moving rocket. Einstein added these two values together despite them being in physically different units (only adding unitless or in-the-same-unit numbers is correct, otherwise unit/number conversion is required). In [P.P.S. to 3.5](#) we highlighted that time-dependent units, like unit of energy Joule, vary in variable time:

There are other "simpler" and more popular derivations of $E = mc^2$ formula, each having its own time related mistake. Here is another [2-minute "proof"](#), which we will correct, for a starter:

$$m = \frac{m_0}{\sqrt{1 - v^2/c^2}} \quad \Longleftarrow \text{Mistake!}$$

$$= m_0(1 - v^2/c^2)^{-\frac{1}{2}}$$

$$= m_0\left(1 + \frac{v^2}{2c^2} + \frac{3v^4}{8c^4} + \ldots\right)$$

$$= m_0 + \frac{m_0 v^2}{2c^2}$$

$$mc^2 = m_0 c^2 + \frac{m_0 v^2}{2}$$

Watch: https://www.youtube.com/watch?v=_dCoW0m4rMs

In 4.1 we discussed Newton's 2nd law **F = m×a**, with mass as measurement for inertial resistance to velocity change. The larger mass **m** is, the less acceleration **a** it gets from a fixed force **F**. The top line in the above screenshot is wrong:

m = m$_0$×D (**Mistake❗**), where **D = 1/sqrt(1–v^2/c^2)** is relativistic time dilation.

In 4.1 we corrected formula **m = m$_0$×D** to formula **m = m$_0$×D^2**. Now,

m = m$_0$×D^2 = m$_0$/(1–v^2/c^2) = m$_0$×(1+ v^2/c^2 + v^4/c^4 + …) = m$_0$ + m$_0$×v^2/c^2 + …

Here we used the formula for geometric series 1/(1–x) = 1 + x + x^2 + …

For v << **c** (v significantly smaller than **c**) we ignore smaller members and keep

m = m$_0$ + m$_0$×v^2/c^2.

Multiply by **c^2**: **m×c^2 = m$_0$×c^2 + m$_0$×v^2**.

Divide by 2: **m×c^2/2 = m$_0$×c^2/2 + m$_0$×v^2/2**.

Thus, **m×c^2/2 accumulates kinetic energy** m$_0$×v^2/2: when kinetic energy is added to m$_0$×c^2/2, it still remains m×c^2/2, with m$_0$ now changed to m. Therefore, m×c^2/2, and not m×c^2, accumulates energy:

E = m×c^2/2 is the total energy of a body!

We have corrected one of the traditional proofs, but let's derive this formula directly from what we have learned about Time Potential in chapter 3:

3.1: Potential = $0.5 \times c^2/D^2$

3.6: $D = 1/\text{sqrt}(1-v^2/c^2)$

By substituting D from 3.6 into 3.1, we get:

Potential = $0.5 \times c^2 \times (1-v^2/c^2) = 0.5 \times c^2 - 0.5 \times v^2$

Potential + $0.5 \times v^2 = 0.5 \times c^2$

By multiplying both sides by mass m, we get:

Potential $\times$ m + $0.5 \times m \times v^2 = 0.5 \times m \times c^2$

Potential energy + Kinetic energy = $m \times c^2/2$

Total energy $E = m \times c^2/2$.

Now, with $E = m \times c^2/2 = m_0 \times D^2 \times c^2/2$, where m_0 and c are constants and only D varies on the right side, we can see clearly that energy E is accumulated in time dilation D, by time in matter m_0 slowing down.

P.S. Does replacing formula $E = m \times c^2$ with $E = m \times c^2/2$ cause major problems for physicists?

Actually, no. Recalibrating constants, units, recalculating results of experiments, etc. happen in physics all the time. The primary usage of formula $E = m \times c^2$ was for determining energy yield in fission and fusion reactions, when on the left side of the energy conservation formula physicists put entry element (like Uranium or Plutonium for fission, or Deuterium and Tritium for fusion) energies for masses M…, and on the right side they put produced-by-fission-or-fusion element energies for masses m… plus energy yield:

$M_1 \times c^2 + M_2 \times c^2 + \ldots = m_1 \times c^2 + m_2 \times c^2 + \ldots +$ energy yield.

And since masses of all participating elements/isotopes and elementary particles were determined beforehand, energy yield was calculated as:

$$\text{energy yield} = \left(\sum M - \sum m\right) \times c^2.$$

Now, with the corrected formula for energy $E = m \times c^2/2$, we have to recalibrate

$$\text{energy yield} = \left(\sum M - \sum m\right) \times c^2/2.$$

Besides that, there could be other implications not directly related to the mass-energy formula. Even for radiation/light energy (where Planck's constant is used) some recalibration might be needed, for example, when energy of an annihilation photon is matched to the energy of an annihilated particle.

Physicists do adjustments/recalibrations all the time, they even do so-called "renormalization" of infinite theoretical values to finite/practical values.

P.P.S. How did such an energy difference go unnoticed?

Simplest example: It was estimated that of 64 kilograms of Uranium used in the bomb that exploded over Hiroshima, only 0.7 grams were converted into energy, the rest became radioactive waste. Could it be an inaccurate assessment? Could it be that 1.4 grams instead of 0.7 grams were converted into energy? Yes. Less than a gram missing of the initial 64 kilograms of Uranium or after-fission various elements / isotopes and elementary particles is hard to measure. And gamma, x-ray, thermal radiation, mechanical energy of the blast etc. – all that is hard to account for accurately, impossible actually. And there were hugely disastrous mistakes / miscalculations acknowledged, especially with H-bomb tests.

As for the "simplest" lab experiments, like electron–positron annihilation, there is still room for revision of measurements, constants, velocities, frequencies, relativistic time dilation effects, adjustments previously made and now taken for granted, etc. As we mentioned in P.P.S. to 3.5, Planck constant is not absolute. But there is more to it: check the next section. No rush, just another refactorization by 0.5.

4.3. E = h×f/2 and Millikan's Test ★★

Let's review Planck's famous formula with his constant h for light/photon energy $E = h \times f$, where f is the frequency of light/photon and $h = 6.626 \times 10^{-34}$ Joule×Sec constant. We have a strong reason for this revision because in the previous section we changed Einstein's mass-energy formula $E = m \times c^2$ to $E = m \times c^2/2$, and because by energy conservation principle, Einstein's and Planck's formulas are linked through electron-positron annihilation, when mass energy of particles is converted into light energy. There are various setups/configurations of electron-positron annihilation, with different numbers of photons at different energies output, but we are interested in the most popular case called para-positronium annihilation with two-photon outcome. The mass-energy of the electron (the same energy and mass the positron has) equals the energy of an annihilation photon. But now with Einstein's formula revised to $E = m \times c^2/2$, we either have to change Planck's formula to $E = h \times f/2$ or change h-value to 3.313×10^{-34} Joule×Sec. But before doing that, we have to review the experiments measuring light energy and h value:

The neatest setup, considered as an icon of experimental physics, is 1912-1915 Millikan's test for photon energy: he measured energy of light by measuring energy of electrons kicked out by photoelectric effect from a metallic plate, and he did that by measuring stop-voltage preventing electrons from reaching another metallic plate:

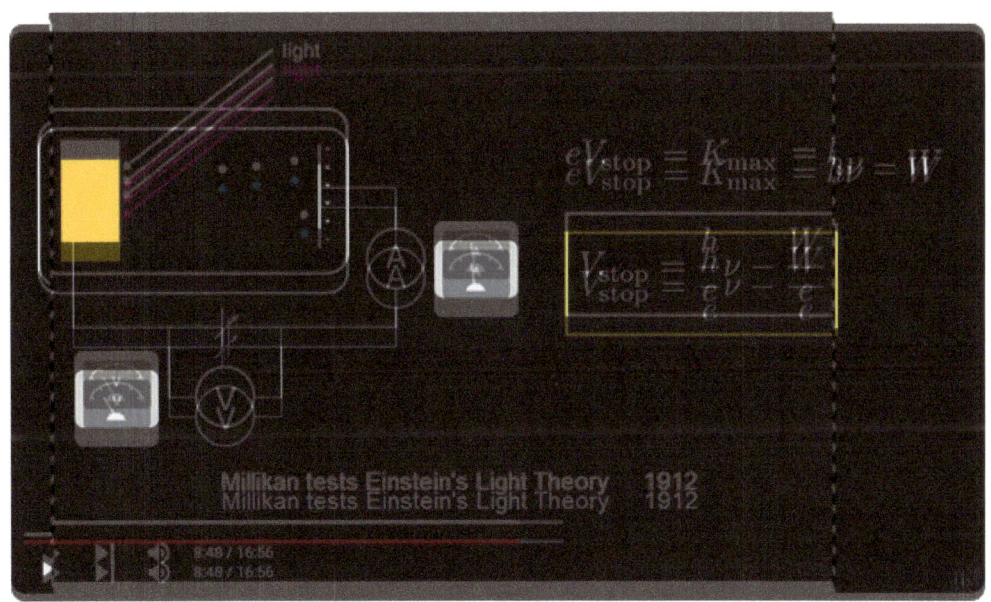

https://www.youtube.com/watch?v=fQzirkrXOxk

For more detailed explanations of Millikan's experiment, please watch the video above and/or the video below.

Electric Current versus Voltage graph below shows that near stop-voltage current is really small: there are very few electrons having such high energy (energy of these few electrons is proportional to the stop-voltage):

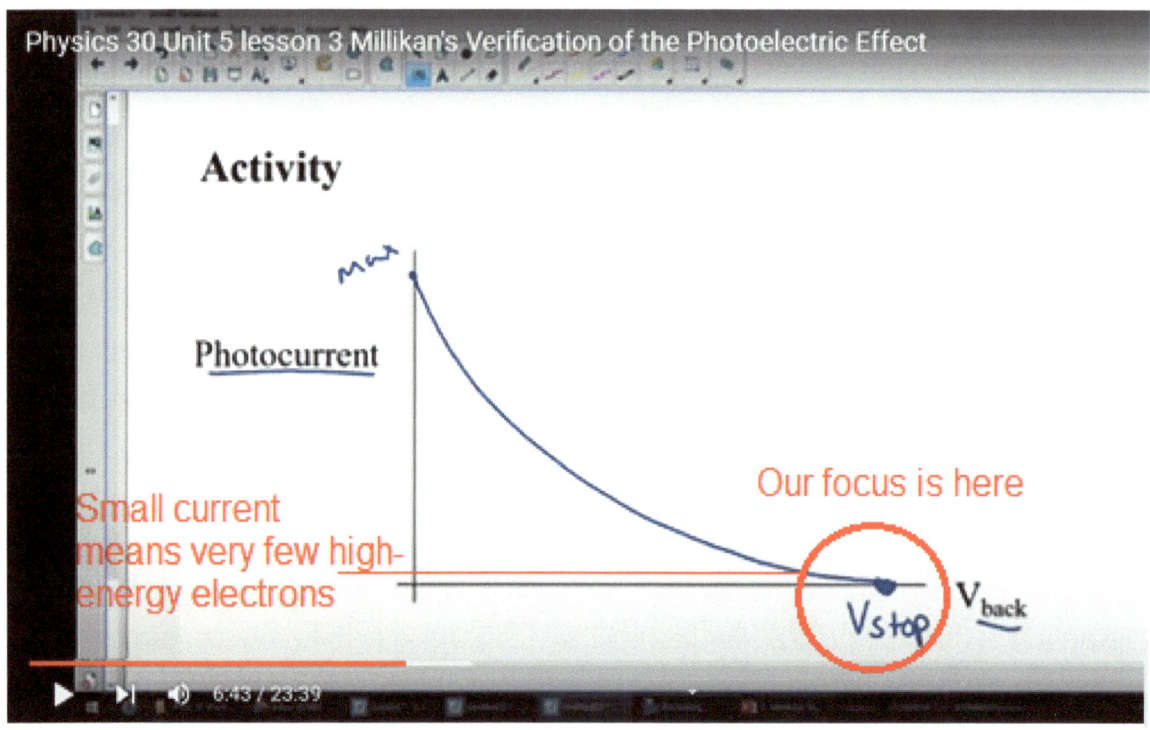

https://youtu.be/4kc3Uv_ATBI?si=5reNSLehr8amWZ3c&t=377

By measuring stop-voltage, Millikan measured maximal kinetic energy (which is $m \times v^2/2$, where **m** is mass of the electron and **v** its velocity) of photoelectrons, thus, he measured **maximum velocity** of these photoelectrons. And the graph clearly shows that the population of such top-velocity electrons is **extremely small**. Millikan assumed that energy of the incident photons equals the energy of the fastest electrons. Is it true or is there twice the difference between these energies? I pay attention to this part specifically, because by-the-factor-of-2 mistakes are quite common in energy and velocity estimates, take for example Feynman's "(Wrong!)" comment in his "43-3 The drift speed" lecture at https://www.feynmanlectures.caltech.edu/I_43.html. Now, let's consider collision at 90° angle between two electrons, initially having the same velocity **v**:

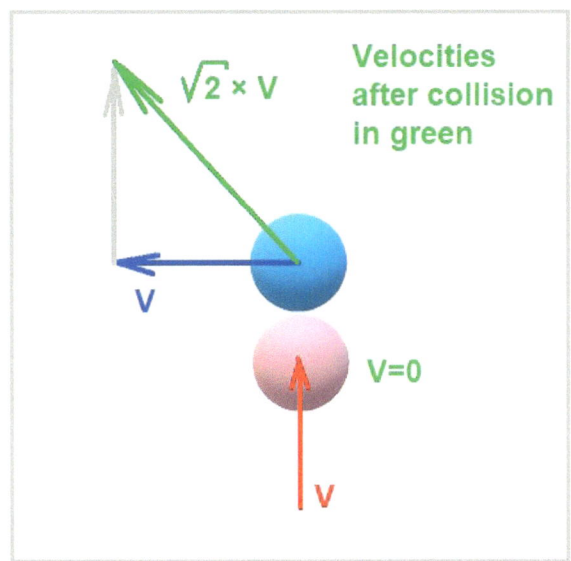

The pink ball stops after collision, and the blue ball gains velocity, and its energy becomes twice higher than before the collision: $\mathbf{m \times (\sqrt{2} \times v)^2 / 2 = m \times v^2 =}$ $\mathbf{2 \times (m \times v^2/2)}$. Both total energy and total momentum are conserved in such elastic collisions. Applying that to photoelectrons: besides the main population of electrons having about the same velocity $\mathbf{v}$ (pointing in various directions), there is a very small population of collided electrons, where some electrons have up to twice the energy of an electron from the main population of photoelectrons. Here is Millikan's mistake: he put an equal sign between photons energy and the maximum of electron's energy, when he should have used only half of the max of electron energy. We just have corrected Planck's photon energy formula to $\mathbf{E = h \times f/2}$, or if you still prefer $\mathbf{E = h \times f}$, then you have to change the value of $\mathbf{h}$ from 6.626×10^{-34} to 3.313×10^{-34} Joule×Sec. But changing the value of $\mathbf{h}$ is more impactful for physics, as you will have to check all formulas in physics that include $\mathbf{h}$.

5. CROSSING BORDER BETWEEN TIMEZONES

5.1. Mechanics by Snell's Law: Escape Velocity

In chapter 2 we discussed that a light beam running in a slow time area, where time dilation is $B > 1$, can penetrate into normal time area (where $B \equiv 1$) only at an angle of incidence $\theta < \arcsin(1/B)$:

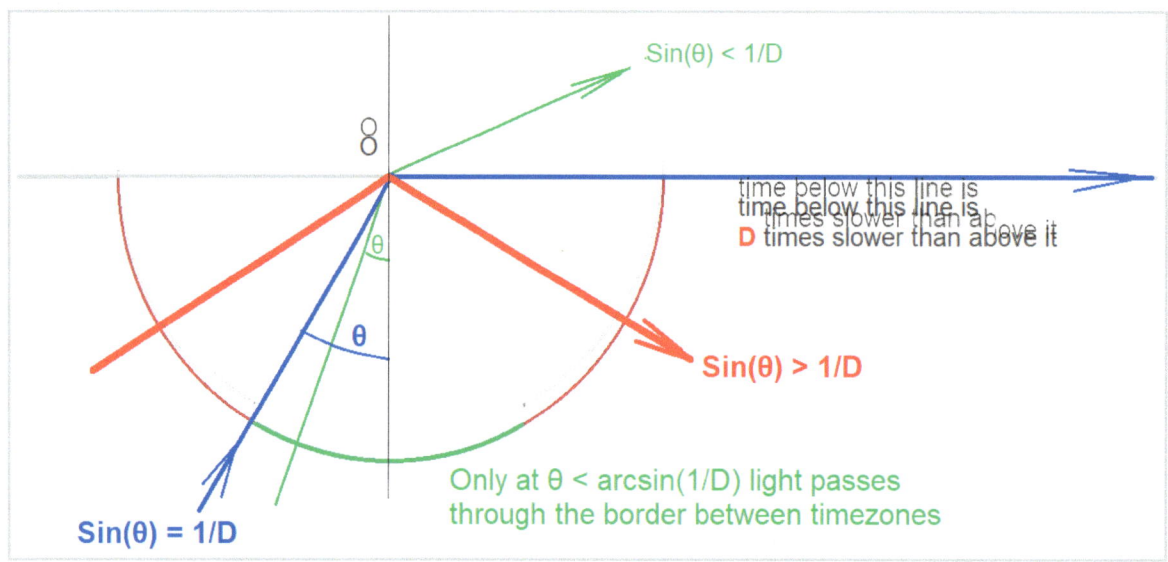

Time Potential is $0.5 \times c^2/B^2$ in the area with time dilation B; and Time Potential is $0.5 \times c^2$ in the area with $B \equiv 1$. An object of mass m needs extra energy $m \times (0.5 \times c^2 = 0.5 \times c^2/B^2)$ to get out from a slower time area. It can get it from its kinetic energy $m \times v^2/2$:

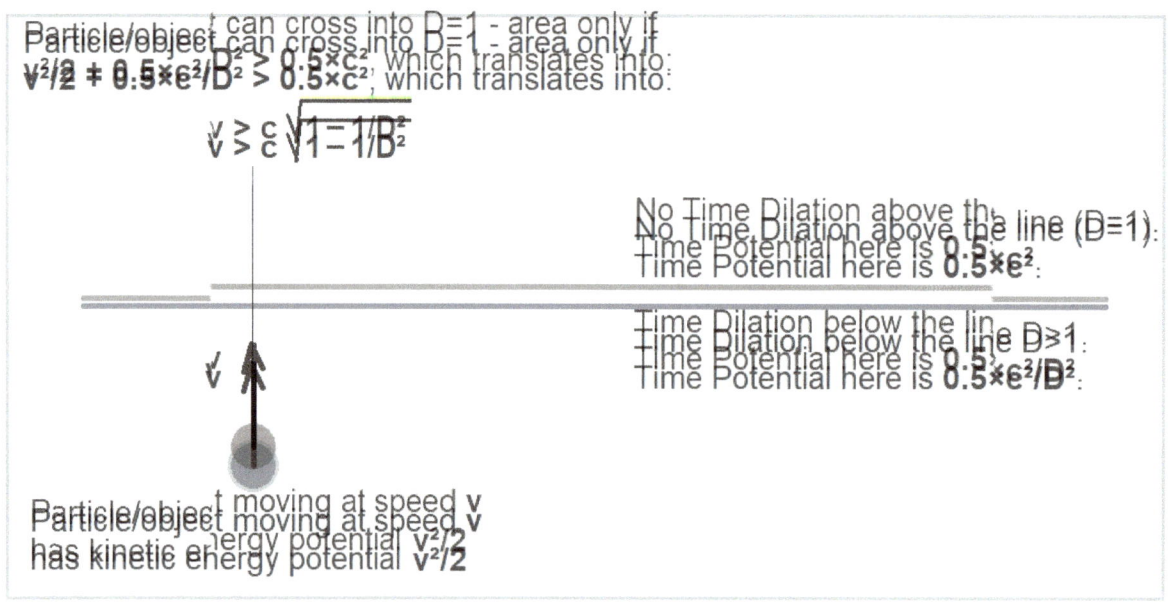

36

In order for an object to cross the border, its velocity **v** should meet the requirement $v > c \times \sqrt{1-1/D^2}$. Even more: the perpendicular to the timezone's border projection of this velocity needs to meet that requirement, because parallel to the border projection of this velocity does not participate in the border crossing.

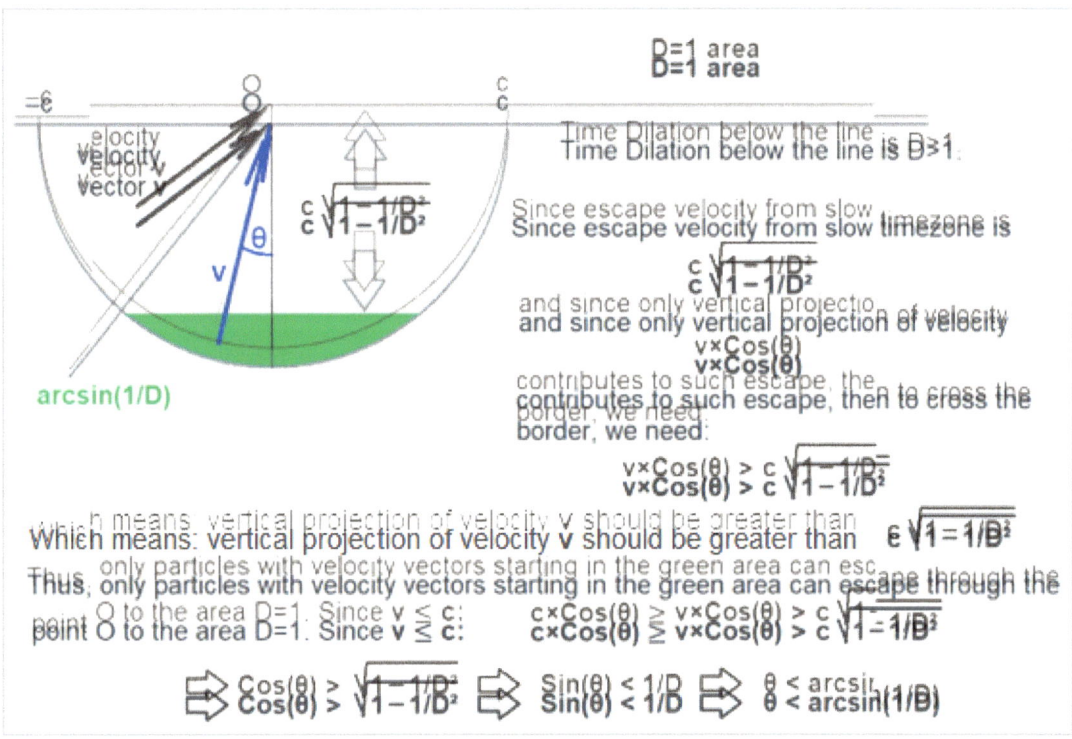

This drawing demonstrates that a particle with a velocity vector, starting in the green cap area of the sphere of radius **c** with a center at the point **O**, can get through point **O**. Actually, we should have started velocity vectors at the object location near or at the point **O**, but then these vectors would span over the border, and that could be misunderstood. Escape velocities (from area with $D \geq 1$) start in the green cap:

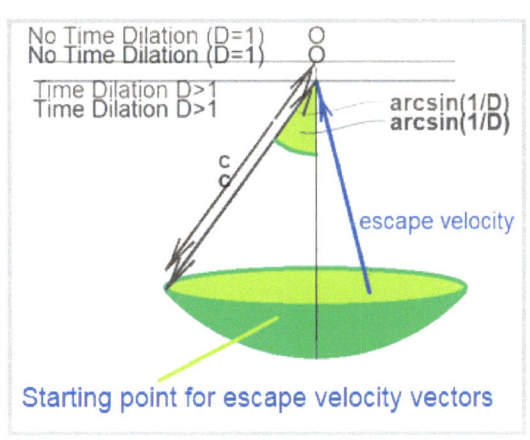

5.2. Strong Force and Half–Life

Light, particles, and objects cannot escape an area with **D = 2** at an incidence angle >30° (because arcsin(1/**D**) = arcsin(1/2) = 30°):

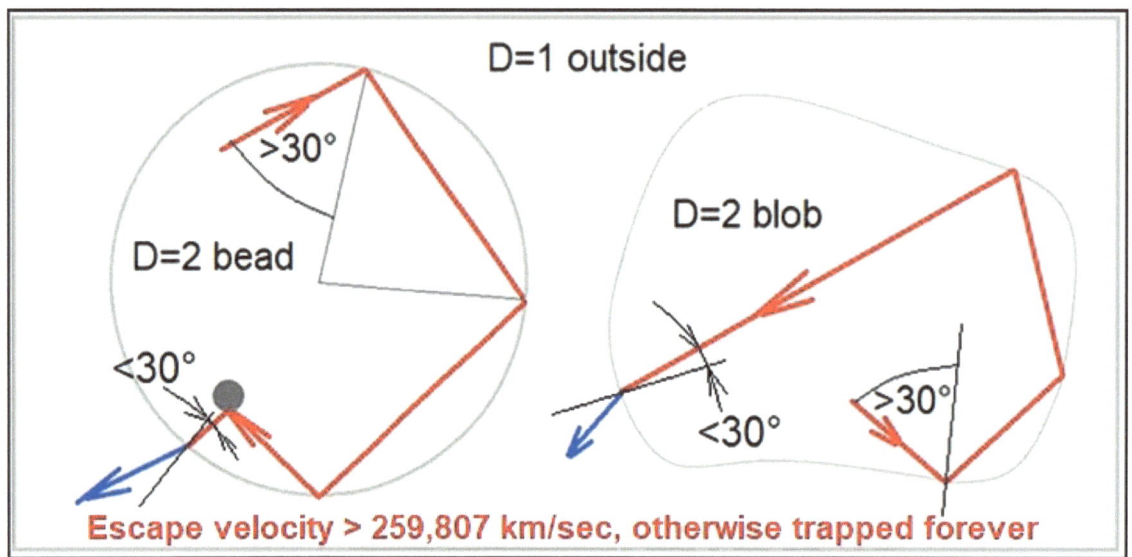

In a spherical bead of twice slower time than outside, light and particles bounce off the surface at the same angle > 30° as the incidence angle, so they are trapped forever, unless something deflects them. From a non-spherical blob, light and particles can escape eventually at some point where incidence angle < 30°. Besides that, if we are talking about a particle, if its velocity < Cos(30°)×**c** = √3 / 2 ×**c** ≈ 0.86×**c** ≈ 259,807 km/sec, it can never escape from any blob with **D = 2**, unless something boosts its velocity up. Here we use sqrt for square root. Time dilation, even at not very big numbers **D**, is a "strong force" capable of containing particles in a nucleus and photons in spherical particles for a long time, even indefinitely, if no severe disturbances/interactions. In particle and nuclear physics, the lifetime of particles and atomic elements is expressed by half–life: it is a time period during which half of the species (of these particles or nuclei) decay. Half–life is inversely related to probability of decay: longer lifetime corresponds to lower probability of decay in a fixed time period. Let's explore what happens to a particle or a nucleus half–life, when time slows down inside them (time dilation factor **D** increases). Intuition suggests that lifetime of such species (which is inversely related to the probability of decay, and decay is about some parts escaping their confinement) is proportional to time dilation **D**, because from the outside perspective, velocities of parts inside the confinement decrease by **D** (see (1.1)), and so is frequency of escape attempts (number of contact points with timezone's border, per outside–observer time unit) decreases. That is only partially correct, because probability of escape at a single contact with the border decreases by **D⁴** as well, thus, combined with the reduction in escape attempts, decay probability decreases by **D⁵**. To prove that, let's bring the last drawing from 5.1 for escape through point **O** velocities:

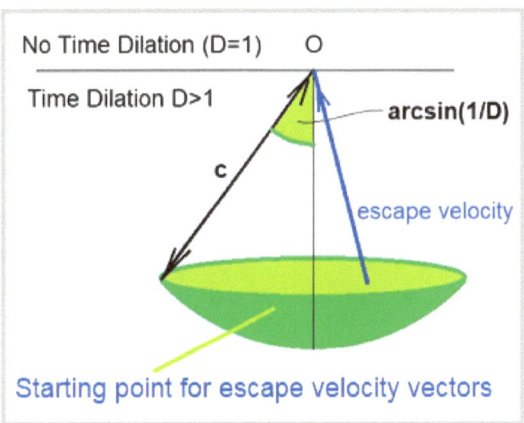

Let's use a known formula for a spherical cap volume:

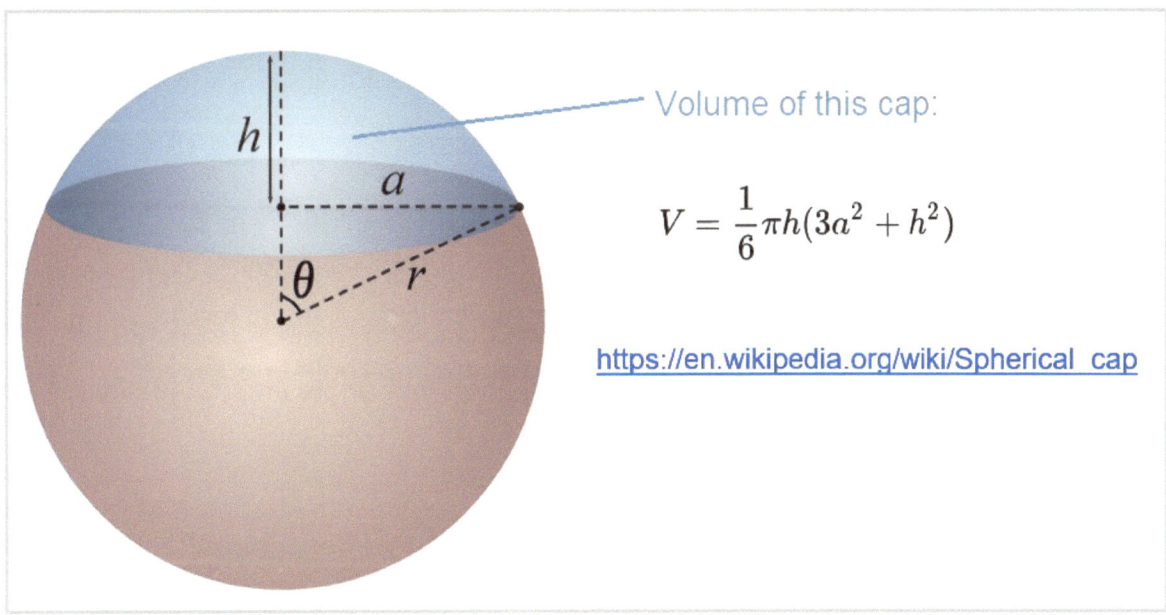

In our case:

r = **c**, θ = arcsin(1/**D**), 1/**D** = a/r = a/**c**, a = **c**/**D**, h = **c**×(1–sqrt(1–1/**D²**))

When B is large, $x \equiv 1/B^2$ is small. We take only the first two members of Tayler series for sqrt(1-x), with $x \equiv 1/B^2$:   sqrt(1-x) $\approx$ 1-0.5x $\equiv$ 1-0.5/B^2

$$h \equiv c \times (1-\text{sqrt}(1-1/B^2)) \approx c \times (1-(1-0.5/B^2)) \approx 0.5 \times c/B^2$$

$$V \approx 1/6 \times \pi \times 0.5 \times c/B^2 \times (3 \times c^2/B^2 + 0.25 \times c^2/B^4) \sim c^3/B^4$$

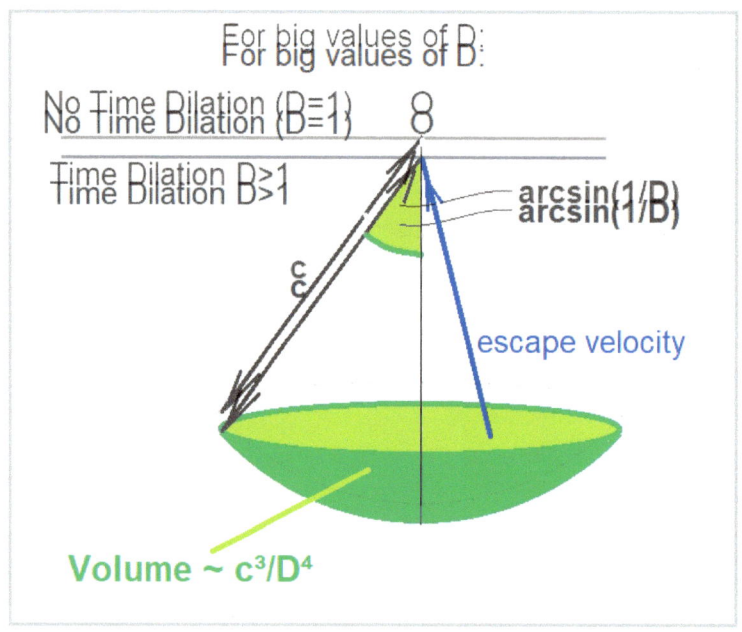

Probability of velocity starting at the green cap (comparing to the whole hemisphere of radius c as 100%, which volume is $2/3 \times \pi \times c^3$) is proportional to

$$(c^3/B^4) / (2/3 \times \pi \times c^3) \sim 1/B^4.$$

As we have mentioned above, combined with decline in frequency of contacting the border by $1/B$, that reduces probability of decay/escape by B^5. In other words, half-life increases by B^5: atom with 10-times-slower atomic clock lives at least 100,000 times longer (if not forever, when parts have no energy or angle enough to escape through a higher and narrower barrier)!

P.S. Read "Time Matters" free eBook for more.

If you are interested in fascinating applications of this physics/math to the real world, cosmology, UFOs etc.; check "Time Matters" (PDF; free on Amazon and Google Books):

www.ingramcontent.com/pod-product-compliance
Lightning Source LLC
Chambersburg PA
CBHW051950210526
45474CB00003B/73